Three Lectures on Low-Dimensional Topology in Kyoto

Lecture Notes on Generalized Heegaard Splittings

Three Lectures on Low-Dimensional Topology in Kyoto

Lecture Notes on Generalized Heegaard Splittings

Martin Scharlemann
University of California, Santa Barbara, USA

Jennifer Schultens
University of California, Davis, USA

Toshio Saito
Joetsu University of Education, Japan

World Scientific

NEW JERSEY · LONDON · SINGAPORE · BEIJING · SHANGHAI · HONG KONG · TAIPEI · CHENNAI · TOKYO

Published by

World Scientific Publishing Co. Pte. Ltd.

5 Toh Tuck Link, Singapore 596224

USA office: 27 Warren Street, Suite 401-402, Hackensack, NJ 07601

UK office: 57 Shelton Street, Covent Garden, London WC2H 9HE

Library of Congress Cataloging-in-Publication Data
Names: Scharlemann, Martin, 1948– | Schultens, Jennifer, 1965– |
    Saito, Toshio (Mathematician)
Title: Lecture notes on generalized Heegaard splittings / by Martin Scharlemann
    (UC Santa Barbara), Jennifer Schultens (UC Davis),
    Toshio Saito (Joetsu University of Education, Japan).
Description: New Jersey : World Scientific, 2016. | Includes bibliographical references and index.
Identifiers: LCCN 2016003719 | ISBN 9789813109117 (hardcover : alk. paper)
Subjects: LCSH: Three-manifolds (Topology) | Topological manifolds. |
    Handlebodies. | Manifolds (Mathematics)
Classification: LCC QA613.2 .S33 2016 | DDC 514/.34--dc23
LC record available at http://lccn.loc.gov/2016003719

British Library Cataloguing-in-Publication Data
A catalogue record for this book is available from the British Library.

Printed in Singapore

Preface

These notes grew out of a lecture series given at RIMS in the summer of 2001. The authors were visiting RIMS in conjunction with the Research Project on Low-Dimensional Topology in the Twenty-First Century. They had been invited by Professor Tsuyoshi Kobayashi. The lecture series was first suggested by Professor Hitoshi Murakami.

The lecture series was aimed at a broad audience that included many graduate students. Its purpose lay in familiarizing the audience with the basics of 3-manifold theory and introducing some topics of current research. The first portion of the lecture series was devoted to standard topics in the theory of 3-manifolds. The middle portion was devoted to a brief study of Heegaard splittings and generalized Heegaard splittings. The latter portion touched on a brand new topic: fork complexes.

During this time Professor Tsuyoshi Kobayashi had raised some interesting questions about the connectivity properties of generalized Heegaard splittings. The latter portion of the lecture series was motivated by these questions. And fork complexes were invented in an effort to illuminate some of the more subtle issues arising in the study of generalized Heegaard splittings.

In the standard schematic diagram for generalized Heegaard splittings, Heegaard splittings are stacked on top of each other in a linear fashion. See Fig. 0.1. This can cause confusion in those cases in which generalized Heegaard splittings possess interesting connectivity properties. In these cases, some of the topological features of the 3-manifold are captured by the connectivity properties of the generalized Heegaard splitting rather than by the Heegaard splittings of submanifolds into which the generalized Heegaard splitting decomposes the 3-manifold. See Fig. 0.2. Fork complexes provide a means of description in this context.

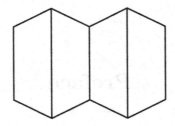

Fig. 0.1   The standard schematic diagram.

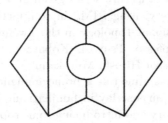

Fig. 0.2   A more informative schematic diagram for a generalized Heegaard splitting for a manifold homeomorphic to (a surface) $\times S^1$.

The authors would like to express their appreciation of the hospitality extended to them during their stay at RIMS. They would also like to thank the many people that made their stay at RIMS delightful, illuminating and productive, most notably Professor Hitoshi Murakami, Professor Tsuyoshi Kobayashi, Professor Jun Murakami, Professor Tomotada Ohtsuki, Professor Kyoji Saito, Professor Makoto Sakuma, Professor Kouki Taniyama and Dr. Yo'av Rieck. Finally, they would like to thank Dr. Ryosuke Yamamoto for drawing the fine pictures in these lecture notes.

M. Scharlemann
J. Schultens
T. Saito

Contents

Chapter 1

Preliminaries

The class of 3-manifolds provides a rich and beautiful world within the field of topology. Many great minds have contributed to our understanding of this world. One of the interesting classical features of 3-manifolds (to emerge towards the middle of the last century) is the fact that the categories DIFF (differentiable or smooth manifolds), TOP (topological manifolds) and PL (piecewise linear or PL manifolds) are equivalent. Standard references for the manifolds and the categories in which they are defined are [do Carmo (1992)], [Guillemin and Pollack (2010)], [Hatcher (1980's)], [Hempel (2004)], [Jaco (1980)], [Massey (1977)], [Milnor (1997)], or [Singer and Thorpe (1976)]. Each category offers a set of tools to study manifolds. In dimension 3, this allows us to choose the set of tools best equipped to solve a particular problem, to illustrate a certain phenomenon and to suit our particular taste. Our choice here will be to work in the TOP and PL categories.

1.1 Manifolds

The central objects under consideration are 3-manifolds.

Definition 1.1.1. A topological space M is called a (topological) 3-*manifold* if it is a second countable Hausdorff space and there exists an open cover $\{U_\alpha\}$ such that U_α is homeomorphic to an open subset of $\mathbb{R}^3$ for any α. Two 3-manifolds are considered equivalent if they are homeomorphic.

Example 1.1.2.

(1) The 3-sphere $\mathbb{S}^3$ is a 3-manifold.
(2) The product $\mathbb{S}^2 \times \mathbb{S}^1$ is a 3-manifold.

We will also consider 3-manifolds with boundary.

Definition 1.1.3. The subset $\mathbb{H}^3 = \{(x, y, z) \in \mathbb{R}^3 : z \geq 0\}$ is called *upper half space*. A topological space M is called a (topological) 3-*manifold with boundary* if it is a second countable Hausdorff space and there exists an open cover $\{U_\alpha\}$ such that U_α is homeomorphic to an open subset of $\mathbb{H}^3$ for any α. Two 3-manifolds with boundary are considered equivalent if they are homeomorphic.

We use the notation $\partial \mathbb{H}^3 = \{(x, y, 0) \in \mathbb{H}^3\}$ and call this subset the *boundary of upper half space*. The subset of M corresponding to such points is called the boundary of M and denoted ∂M. It is an exercise in [Guillemin and Pollack (2010), Chapter 1] that the boundary of M is well-defined and is a closed 2-manifold. (Recall that a manifold is said to be closed if it is compact and without boundary.)

Example 1.1.4.

(1) The 3-ball $\mathbb{B}^3$ is a 3-manifold with boundary. Its boundary is the 2-sphere $\mathbb{S}^2$.
(2) The solid torus, *i.e.*, the product $\mathbb{B}^2 \times \mathbb{S}^1$ is a 3-manifold with boundary. Its boundary is the torus $\mathbb{T}^2 = \mathbb{S}^1 \times \mathbb{S}^1$.

Exercise 1.1.5. Describe all self-homeomorphisms of the torus.

Exercise 1.1.6. Show that the result of identifying two solid tori via a self-homeomorphism of the boundary tori is a 3-manifold. A 3-manifold obtained in this way is called a *lens space*.

Remark 1.1.7. Lens spaces are completely classified by Reidemeister. See [Reidemeister (1935)]. See also Remark 2.2.7 in the next chapter.

1.2   Simplicial complexes

Recall that a collection $\{v_0, v_1, \ldots, v_n\}$ of $n + 1$ points in $\mathbb{R}^N$ is said to be *geometrically independent* if the vectors $\{v_1 - v_0, \ldots, v_n - v_0\}$ are linearly independent.

Definition 1.2.1. Let $\{v_0, v_1, \ldots, v_n\}$ be a geometrically independent collection of $n + 1$ points in $\mathbb{R}^N$. An *n-simplex* σ^n, which is also denoted $\langle v_0, v_1, \ldots, v_n \rangle$, spanned by these points is defined to be

$$\sigma^n = \left\{ \sum_{i=0}^{n} t_i v_i \in \mathbb{R}^N : \sum_{i=0}^{n} t_i = 1, \ t_i \geq 0 \text{ for each } i \right\}.$$

A simplex spanned by a sub-collection of $\{v_0, v_1, \ldots, v_n\}$ is called a *face* of σ^n. An m-face of σ^n is a face of σ^n with m vertices. The *barycenter* of σ^n is defined to be $\dfrac{1}{n+1} \sum_{i=0}^{n} v_i$.

Remark 1.2.2. The empty set is a face of any simplex.

Example 1.2.3. Low-dimensional simplices are shown in Fig. 1.1. A 0-simplex is a point and is also called a *vertex*, a 1-simplex is a line segment and is also called an *edge*, a 2-simplex is a triangle, and a 3-simplex is a tetrahedron.

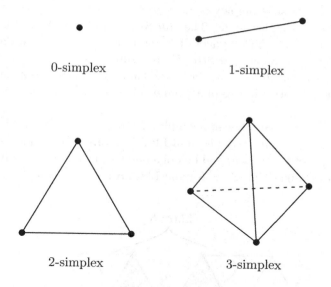

0-simplex

1-simplex

2-simplex

3-simplex

Fig. 1.1 Low-dimensional simplices.

Exercise 1.2.4. Show that the total number of faces of an n-simplex is 2^{n+1}.

Definition 1.2.5. A *simplicial complex* is a finite set K of simplices in $\mathbb{R}^N$ which satisfies the following:

(1) every face of any simplex in K is a member of K, and
(2) the intersection of any two simplices is a member of K.

The union of all the simplices of K in $\mathbb{R}^N$ is called the *(underlying) poly-hedron* of K and is denoted by $|K|$.

Exercise 1.2.6. Draw the two dimensional polyhedron of the following simplicial complex:

$$\{\emptyset, \langle v_0 \rangle, \langle v_1 \rangle, \langle v_2 \rangle, \langle v_3 \rangle, \langle v_4 \rangle, \langle v_5 \rangle, \langle v_0, v_1 \rangle, \langle v_1, v_2 \rangle,$$
$$\langle v_2, v_0 \rangle, \langle v_2, v_3 \rangle, \langle v_3, v_4 \rangle, \langle v_4, v_5 \rangle, \langle v_5, v_3 \rangle, \langle v_0, v_1, v_2 \rangle\}.$$

Exercise 1.2.7. A *sub-complex* of K is a simplicial complex that is a sub-collection of K.

(1) Show that an *m-skeleton*, *i.e.*, the collection of all the m'-simplices with $m' \le m$, is a sub-complex of K.
(2) Let σ be a simplex in K. The *star* $\mathrm{St}(\sigma; K)$ of σ is defined to be the collection of all the simplices of K intersecting σ and of all their faces (cf. Fig. 1.2). Show that $\mathrm{St}(\sigma; K)$ is a sub-complex of K.
(3) Show that the *link* $\mathrm{Lk}(\sigma; K)$ of σ, which is the collection of all the simplices in $\mathrm{St}(\sigma; K)$ disjoint from σ (cf. Fig. 1.2), is a sub-complex of K.
(4) The star and link of a sub-complex K' of K are similarly defined as above. Show that $\mathrm{St}(K'; K)$ and $\mathrm{Lk}(K'; K)$ are sub-complexes of K.
(5) More generally, the star and link of a point x in $|K|$ are similarly defined as above. Show that $\mathrm{St}(x; K)$ and $\mathrm{Lk}(x; K)$ are sub-complexes of K.

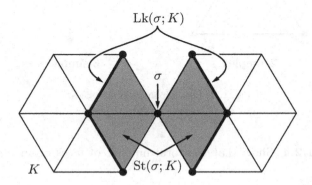

Fig. 1.2   The star and link of a simplex σ in a simplicial complex K.

Definition 1.2.8. A simplicial complex K' is called a *subdivision* of a simplicial complex of K if $|K'| = |K|$ and every simplex in K' is contained in a simplex in K.

Example 1.2.9. One very useful subdivision is called the *barycentric subdivision*. It is described as follows:

The barycentric subdivision, denoted $\mathrm{Sd}(\sigma^n)$, of an n-simplex σ^n is defined inductively. The barycentric subdivision $\mathrm{Sd}(\sigma^0)$ of σ^0 is defined to be σ^0 itself. (Note that $\mathrm{Sd}(\sigma^0) = \sigma^0$ is also equal to its own barycenter.)

Next assume that we have defined the barycentric subdivision $\mathrm{Sd}(\sigma^{n-1})$ of σ^{n-1} as a simplicial complex. Take $\mathrm{Sd}'(\sigma^n)$ to be the union of all simplices in the barycentric subdivisions of all the $(n-1)$-faces of σ^n. Further, for every simplex $\tau \in \mathrm{Sd}'(\sigma^n)$, we take the simplex $\widehat{\tau}$ spanned by all the vertices of τ together with the barycenter σ_c^n of σ^n. (Note that if τ is a simplex of dimension m, then $\widehat{\tau}$ is a simplicial complex of dimension $m+1$.) We thus obtain the set $\mathrm{Sd}''(\sigma^n) = \{\widehat{\tau} : \tau \in \mathrm{Sd}'(\sigma^n)\}$.

Then the barycentric subdivision $\mathrm{Sd}(\sigma^n)$ of σ^n is defined to be

$$\mathrm{Sd}(\sigma^n) = \{\sigma_c^n\} \cup \mathrm{Sd}'(\sigma^n) \cup \mathrm{Sd}''(\sigma^n).$$

The barycentric subdivision, denoted by $\mathrm{Sd}(K)$, of a simplicial complex K is formed by the barycentric subdivisions of all the simplices of K.

Definition 1.2.10. Let K be a simplicial complex and X a sub-complex of K. Let $\mathrm{Sd}^{(2)}(K)$ be the second barycentric subdivision of K, *i.e.*, the barycentric subdivision of $\mathrm{Sd}(K)$. The *regular neighborhood* of X in K, denoted by $\eta(X; K)$, is defined to be $\mathrm{St}(X; \mathrm{Sd}^{(2)}(K))$ (cf. Fig. 1.3).

1.3 PL manifolds

Definition 1.3.1. Let K and L be simplicial complexes. Polyhedra $|K|$ and $|L|$ are said to be *piecewise linearly (PL) homeomorphic* if there are subdivisions K' and L' of K and L respectively such that there is a face-preserving bijection between K' and L'.

Definition 1.3.2. Let σ^n be an n-simplex and $\partial\sigma^n$ a subset of its faces other than the whole set σ^n. The polyhedron $|\sigma^n|$ ($|\partial\sigma^n|$ resp.) is called a *PL n-ball* (a *PL $(n-1)$-sphere* resp.).

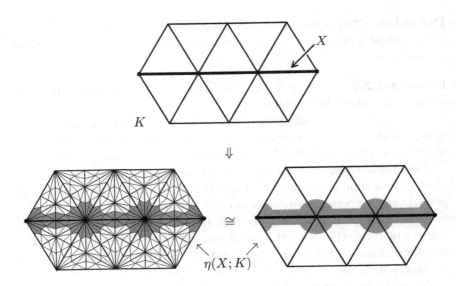

Fig. 1.3   A two dimensional version of a regular neighborhood of X.

Definition 1.3.3. Let M be a manifold. A *triangulation* of M is a pair consisting of a simplicial complex K and a homeomorphism $t : M \to |K|$.

Definition 1.3.4. An n-manifold M is called a *PL n-manifold* if there is a triangulation (K, t) of M such that for each point x in M, the link of $t(x)$ in $|K|$ is PL homeomorphic to a PL $(n-1)$-sphere or to a PL $(n-1)$-ball.

Remark 1.3.5. The union of points $x \in M$ satisfying the latter condition in Definition 1.3.4 is the boundary ∂M of M.

We now focus on PL 3-manifolds which are central objects in this book.

Example 1.3.6.

(1) The 3-ball $\mathbb{B}^3$ is the simplest PL 3-manifold in a sense that $\mathbb{B}^3$ is homeomorphic to a 3-simplex.
(2) The 3-sphere $\mathbb{S}^3$ is also a 3-manifold. Since $\mathbb{S}^3$ is homeomorphic to the boundary of a 4-ball which is homeomorphic to a 4-simplex, we see that $\mathbb{S}^3$ is a union of five 3-simplices. It is easy to show that this gives a triangulation of $\mathbb{S}^3$.

Exercise 1.3.7. Show that the following 3-manifolds are PL 3-manifolds.

(1) The solid torus $\mathbb{B}^2 \times \mathbb{S}^1$.
(2) The product $\mathbb{S}^2 \times \mathbb{S}^1$.
(3) The lens spaces.

Definition 1.3.8. Let M be a PL manifold. A sub-manifold N of M is said to be *properly embedded* in M if the interior of N is contained in that of M and ∂N is contained in ∂M.

Example 1.3.9.

(1) A disk $\mathbb{B}^2 \times \{*\}$ in $\mathbb{B}^2 \times \mathbb{S}^1$ is properly embedded.
(2) Let α be a simple closed curve in the torus $\mathbb{T}^2$. Then the annulus $\alpha \times [0,1]$ in $\mathbb{T}^2 \times [0,1]$ is properly embedded.

We state the following propositions on regular neighborhoods without proofs.

Proposition 1.3.10. *If X is a PL 1-manifold properly embedded in a PL 3-manifold M (namely, $X \cap \partial M = \partial X$), then $\eta(X; M)$ is homeomorphic to $X \times \mathbb{B}^2$, where X is identified with $X \times \{a \text{ center of } \mathbb{B}^2\}$ and $\eta(X; M) \cap \partial M$ is identified with $\partial X \times \mathbb{B}^2$ (cf. Fig. 1.4).*

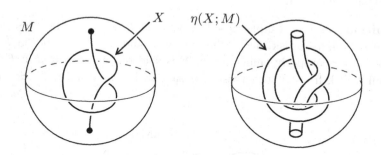

Fig. 1.4   A regular neighborhood of a PL 1-manifold X in a PL 3-manifold M is homeomorphic to $X \times \mathbb{B}^2$.

Proposition 1.3.11. *Suppose that a PL 3-manifold M is orientable. If X is an orientable PL 2-manifold properly embedded in M (namely, $X \cap \partial M = \partial X$), then $\eta(X; M)$ is homeomorphic to $X \times [0,1]$, where X is identified with $X \times \{1/2\}$ and $\eta(X; M) \cap \partial M$ is identified with $\partial X \times [0,1]$.*

One can ask what manifolds are PL manifolds. It is easy to see that every compact 1-manifold is a PL 1-manifold. It was proved by Radó in 1925 that every compact 2-manifold is a PL 2-manifold (cf. [Radó (1925)]). As mentioned at the beginning of this chapter, the same is true for compact 3-manifolds. This was first proved by Moise. See [Moise (1952)].

Theorem 1.3.12 ([Moise (1952)]). *Every compact 3-manifold is a PL 3-manifold.*

We, however, note that there are manifolds of dimension greater than three which are not PL manifolds. Higher dimensional case was done by Kirby and Siebenmann. See [Kirby and Siebenmann (1969)]. Works of Freedman (cf. [Freedman (1982)]) and Casson (cf. [Akbulut and McCarthy (1990)]) gave an example of a 4-manifold which is not a PL manifold.

In the remainder of this book, we work in the PL category unless otherwise specified.

1.4   Fundamental definitions

By the term *surface*, we will mean a connected compact 2-manifold. Throughout this section, let F be a surface and M a connected compact orientable 3-manifold.

Definition 1.4.1. A simple closed curve α in F is said to be *inessential* in F if α bounds a disk in F, otherwise α is said to be *essential* in F. A (simple) arc γ properly embedded in F is said to be *inessential* in F if γ cuts off a disk from F, otherwise γ is said to be *essential* in F. See Fig. 1.5.

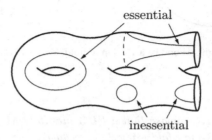

essential

inessential

Fig. 1.5

Definition 1.4.2. A disk D properly embedded in M is said to be *inessential* in M if D cuts off a 3-ball from M, otherwise D is said to be *essential* in M. A 2-sphere P properly embedded in M is said to be *inessential* in M if P bounds a 3-ball in M, otherwise P is said to be *essential* in M.

Example 1.4.3.

(1) The disk described in (1) of Example 1.3.9 is essential.
(2) The 2-sphere $\mathbb{S}^2 \times \{1/2\}$ in $\mathbb{S}^2 \times [0,1]$ is essential.

The following terminology provides the necessary framework in which to discuss surfaces.

Definition 1.4.4. Suppose that F is properly embedded in M.

(1) We say that F is *∂-parallel* in M if F cuts off a 3-manifold homeomorphic to $F \times [0,1]$ from M.
(2) We say that F is *compressible* in M if there is a disk $D \subset M$ such that $D \cap F = \partial D$ and ∂D is an essential simple closed curve in F. Such a disk D is called a *compressing disk*.
(3) We say that F is *incompressible* in M if F is not compressible in M.
(4) The surface F is *∂-compressible* in M if there is a disk $\delta \subset M$ such that $\delta \cap F$ is an arc which is essential in F, say γ, in F and that $\delta \cap \partial M$ is an arc, say γ', with $\gamma' \cup \gamma = \partial \delta$. Such a disk δ is called a *∂-compressing disk*.
(5) We say that F is *∂-incompressible* in M if F is not *∂*-compressible in M.
(6) Suppose that F is homeomorphic neither to a disk nor to a 2-sphere. Then the surface F is said to be *essential* in M if F is incompressible in M and is not *∂*-parallel in M.

Example 1.4.5.

(1) An annulus properly embedded in a solid torus as illustrated in the left of Fig. 1.6 is *∂*-compressible and is also *∂*-parallel.
(2) A Möbius band properly embedded in a solid torus as illustrated in the right of Fig. 1.6 is *∂*-compressible but is not *∂*-parallel.
(3) Let α be an essential closed curve in the torus $\mathbb{T}^2$. Then the annulus $\alpha \times [0,1]$ in $\mathbb{T}^2 \times [0,1]$ is essential (and hence incompressible).

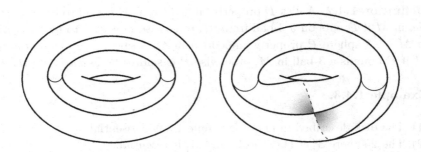

Fig. 1.6   The annulus (left) is ∂-parallel in a solid torus. The Möbius band (right) is not ∂-parallel.

The following terminology allows us to isolate the 3-manifolds that form the basic building blocks for all 3-manifolds.

Definition 1.4.6.

(1) A 3-manifold M is said to be *reducible* if there is a 2-sphere in M which does not bound a 3-ball in M. Such a 2-sphere is called a *reducing 2-sphere* of M. A 3-manifold M is said to be *irreducible* if M is not reducible.

(2) A 3-manifold M is said to be *∂-reducible* if there is a disk properly embedded in M whose boundary is essential in ∂M. Such a disk is called a *∂-reducing disk.*

Example 1.4.7.

(1) It follows from Schoenflies Theorem, which implies that every 2-sphere in $\mathbb{S}^3$ cuts it into two 3-balls, that $\mathbb{S}^3$ is irreducible. See [Schoenflies (1906)].

(2) The solid torus $\mathbb{B}^2 \times \mathbb{S}^1$ is ∂-reducible.

Suppose that a 3-manifold M is reducible and let P be its reducing 2-sphere. If P is separating in M, then P cuts M into two manifolds, say M_1' and M_2'. For each $i = 1, 2$, let M_i be a 3-manifold obtained from M_i' by capping off a 3-ball along the copy of P. We say that M is obtained by the *connected sum*, denoted $M_1 \# M_2$, of M_1 and M_2. A connected sum of $M_1 \# M_2$ is said to be *trivial* if M_1 or M_2 is homeomorphic to $\mathbb{S}^3$. We say a 3-manifold $M (\not\cong \mathbb{S}^3)$ is *prime* if M does not admit a non-trivial connected sum.

Example 1.4.8.

(1) $\mathbb{S}^2 \times [0,1] = \mathbb{B}^3 \# \mathbb{B}^3$.

(2) $\mathbb{S}^2 \times \mathbb{S}^1$ is reducible and is also prime.

Exercise 1.4.9. Let M be a reducible 3-manifold and P its reducing 2-sphere. Show that if P is non-separating in M, then $M \cong M' \# \mathbb{S}^2 \times \mathbb{S}^1$ for some 3-manifold M'.

Chapter 2

Definition and examples of Heegaard splittings

In this chapter we discuss Heegaard splittings. Conceived by Poul Heegaard in his dissertation in 1898, inspired by a trip to Göttingen and the influence of Felix Klein, Heegaard splittings originally provided a framework in which to state a counterexample to Poincaré's original formulation of duality. (Thus forcing Poincaré back into topology.) The motivation for studying Heegaad splittings in recent decades lies in the natural decomposition of 3-manifolds they entail. This decomposition into "symmetric" pieces allows us to visualize many 3-manifolds. We provide several examples to illustrate the type of visualization they allow.

2.1 Definitions

Recall that $\mathbb{B}^n$ denotes the n-ball, *i.e.*, the n-dimensional ball.

Definition 2.1.1. For each integer m with $0 \le m \le 3$, a 3-dimensional *m-handle* is the product $\mathbb{B}^m \times \mathbb{B}^{3-m}$. Its sub-manifold $\mathbb{B}^m \times \{c\}$ is called the *core* of the m-handle, and $\{c\} \times \mathbb{B}^{3-m}$ is called the *co-core* of the m-handle, where c is the center of the ball (cf. Fig. 2.1). The product of the boundary of the core of the m-handle, which is homeomorphic to $\mathbb{S}^{m-1}$, and $\mathbb{B}^{3-m}$ is called the *attaching sphere*.

Definition 2.1.2. A 3-manifold C is called a *compression body* if there exists a closed orientable surface F such that C is obtained from $F \times [0,1]$ by attaching 2-handles along pairwise disjoint loops in $F \times \{1\}$ and filling in resulting 2-sphere boundary components with 3-handles (cf. Fig. 2.2). We denote $F \times \{0\}$ by $\partial_+ C$ and $\partial C \setminus \partial_+ C$ by $\partial_- C$. A compression body C is called a *handlebody* if $\partial_- C = \emptyset$. A compression body C is said to be *trivial* if C is homeomorphic to $F \times [0,1]$.

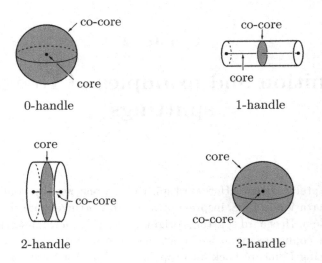

Fig. 2.1   The Core and co-core of a handle.

Definition 2.1.3. For a compression body C, an essential disk in C is called a *meridian disk* of C. A union Δ of pairwise disjoint meridian disks of C is called a *complete meridian system* if the manifold obtained from C by cutting along Δ is the union of $\partial_- C \times [0, 1]$ and (possibly empty) 3-balls. A complete meridian system Δ of C is *minimal* if the number of the components of Δ is minimal among all complete meridian systems of C.

Remark 2.1.4. The following properties are known for compression bodies.

(1) A compression body C is reducible if and only if $\partial_- C$ contains a 2-sphere component.
(2) A minimal complete meridian system Δ of a compression body C cuts C into $\partial_- C \times [0, 1]$ if $\partial_- C \neq \emptyset$, and Δ cuts C into a single 3-ball if $\partial_- C = \emptyset$, *i.e.*, when C is a handlebody.
(3) Extending the cores of the 2-handles, which are in the definition of the compression body C, through $F \times [0, 1]$ vertically, we obtain a complete meridian system Δ of C such that the manifold obtained by cutting C along Δ is homeomorphic to a union of $\partial_- C \times [0, 1]$ and some (possibly

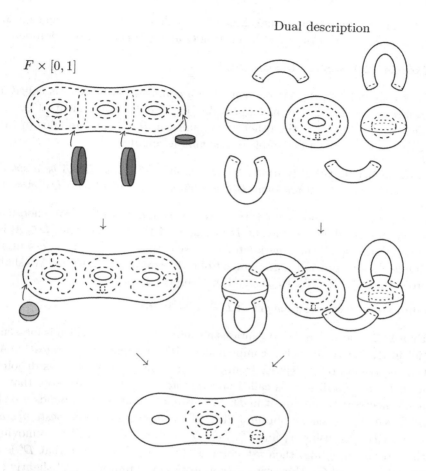

Fig. 2.2   Constructing a compression body and its dual description.

empty) 3-balls. This gives a *dual description* of a compression body. That is, a compression body C is obtained from $\partial_- C \times [0, 1]$ and some (possibly empty) 3-balls by attaching 1-handles to $\partial_- C \times \{1\}$ and to the boundary of the 3-balls (cf. Fig. 2.2).

(4) For any compression body C, $\partial_- C$ is incompressible in C.

(5) Let C and C' be compression bodies. Suppose that C'' is obtained from C and C' by identifying a component of $\partial_- C$ and of $\partial_+ C'$. Then C'' is a compression body.

(6) Let D be a meridian disk of a compression body C. Then there is a

complete meridian system Δ of C such that D is a component of Δ. Any component obtained by cutting C along D is a compression body.

Exercise 2.1.5. Show Remark 2.1.4.

An annulus A properly embedded in a compression body C is called a *spanning annulus* if A is incompressible in C and a component of ∂A is contained in $\partial_+ C$ and the other is contained in $\partial_- C$. The annulus in (3) of Example 1.4.5 is an example of a spanning annulus.

Lemma 2.1.6. *Let C be a non-trivial compression body. Let A be a spanning annulus in C. Then there is a meridian disk D of C with $D \cap A = \emptyset$.*

Proof. Since C is non-trivial, there is a meridian disk of C. We choose a meridian disk D of C such that D intersects A transversely and $|D \cap A|$ is minimal among all such meridian disks. Note that $A \cap \partial_- C$ is an essential simple closed curve in the component of $\partial_- C$ containing $A \cap \partial_- C$. We shall prove that $D \cap A = \emptyset$. To this end, we suppose $D \cap A \neq \emptyset$.

Claim 1. Every component of $D \cap A$ is an arc.

Proof. Suppose that $D \cap A$ contains a simple closed curve which is inessential in A. Let α be such a component of $D \cap A$ which is *innermost* in A, that is, α cuts off a disk δ_α from A so that the interior of δ_α is disjoint from D. Such a disk δ_α is called an *innermost disk* for α. We note that α is not necessarily innermost in D. We also note that α also bounds a disk in D, say δ'_α. Then we obtain a disk D' by applying *cut and paste operation* on D with using δ_α and δ'_α, i.e., D' is obtained from D by removing the interior of δ'_α and then attaching δ_α (cf. Fig. 2.3). Note that D' is a meridian disk of C. Moreover, we can isotope the interior of D' slightly so that $|D' \cap A| < |D \cap A|$, a contradiction. (Such an argument as above is called an *innermost disk argument.*)

Hence if $D \cap A$ contains a simple closed curve, we may assume that the curve is essential in A. Let α' be such a component of $D \cap A$ which is innermost in D, and let $\delta_{\alpha'}$ be the innermost disk in D with $\partial \delta_{\alpha'} = \alpha'$. Then α' cuts A into two annuli, and let A' be the component obtained by cutting A along α' such that A' is adjacent to $\partial_- C$. Set $D'' = A' \cup \delta_{\alpha'}$. Then $D''(\subset C)$ is a compressing disk of $\partial_- C$, contradicting (4) of Remark 2.1.4. Hence we have Claim 1. □

Claim 2. There are no arc components of $D \cap A$.

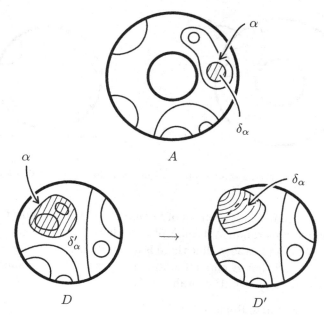

Fig. 2.3   A disk D' obtained from D by removing the interior of δ'_α and then by attaching δ_α along α.

Proof. Suppose that there is an arc component of $D \cap A$. Note that $\partial D \subset \partial_+ C$. Hence we may assume that each component of $D \cap A$ is an inessential arc in A such that both endpoints are contained in $\partial_+ C$. Let γ be an arc component of $D \cap A$ which is *outermost* in A, that is, γ cuts off a disk δ_γ from A such that the interior of δ_γ is disjoint from D. Such a disk δ_γ is called an *outermost disk* for γ. Note that γ cuts D into two disks $\bar{\delta}_\gamma$ and $\bar{\delta}'_\gamma$ (cf. Fig. 2.4).

   If both $\bar{\delta}_\gamma \cup \delta_\gamma$ and $\bar{\delta}'_\gamma \cup \delta_\gamma$ are inessential in C, then D is also inessential in C, a contradiction. Hence we may assume that $\bar{D} = \bar{\delta}_\gamma \cup \delta_\gamma$ is essential in C. Then we can isotope $\bar{D}$ slightly so that $|\bar{D} \cap A| < |D \cap A|$, a contradiction. We therefore have Claim 2. (Such an argument as above is called an *outermost disk argument*.)   □

   It follows from Claims 1 and 2 that $D \cap A = \emptyset$, and this completes the proof of Lemma 2.1.6.   □

Remark 2.1.7. Let A be a spanning annulus in a non-trivial compression body C.

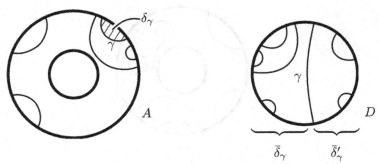

Fig. 2.4   An outermost arc γ of $D \cap A$ in A and its outermost disk δ_γ.

(1) Using an argument in the proof of Lemma 2.1.6, we can show that there
    is a complete meridian system Δ of C with $\Delta \cap A = \emptyset$.
(2) It follows from (1) above that there is a meridian disk E of C such that
    $E \cap A = \emptyset$ and that E cuts off a 3-manifold which is homeomorphic to
    (a closed surface) $\times [0,1]$ containing A.

Exercise 2.1.8. Show Remark 2.1.7.

Let $\bar{\alpha} = \alpha_1 \cup \cdots \cup \alpha_p$ be a union of pairwise disjoint arcs in a compression
body C. We say that $\bar{\alpha}$ is *vertical* if there is a union of pairwise disjoint
spanning annuli $A_1 \cup \cdots \cup A_p$ in C such that $\alpha_i \cap A_j = \emptyset$ $(i \neq j)$ and α_i is
an essential arc properly embedded in A_i $(i = 1, 2, \ldots, p)$.

Lemma 2.1.9. *Let $\bar{\alpha} = \alpha_1 \cup \cdots \cup \alpha_p$ be as above. Suppose that $\bar{\alpha}$ is
vertical in C. Let D be a meridian disk of C. Then there is a meridian
disk D' of C with $D' \cap \bar{\alpha} = \emptyset$ which is obtained by cut-and-paste operation
on D. Particularly, if C is irreducible, then D is ambient isotopic such that
$D \cap \bar{\alpha} = \emptyset$.*

Proof. Let $\bar{A} = A_1 \cup \cdots \cup A_p$ be a union of annuli for $\bar{\alpha}$ as above. Using
innermost disk arguments, we see that there is a meridian disk D' such that
no components of $D' \cap \bar{A}$ are simple closed curves which are inessential in
$\bar{A}$. We remark that D' is ambient isotopic to D if C is irreducible. Note
that each component of $\bar{A}$ is incompressible in C. Hence no components
of $D' \cap \bar{A}$ are simple closed curves which are essential in $\bar{A}$. Hence each
component of $D' \cap \bar{A}$ is an arc; moreover since ∂D is contained in $\partial_+ C$, the
endpoints of the arc components of $D' \cap \bar{A}$ are contained in $\partial_+ C \cap \bar{A}$. Then
it is easy to see that there exists an arc $\beta_i (\subset A_i)$ such that β_i is essential
in A_i and $\beta_i \cap D' = \emptyset$. Take an ambient isotopy h_t $(0 \leq t \leq 1)$ of C such

that $h_0(\beta_i) = \beta_i$, $h_t(\bar{A}) = \bar{A}$ and $h_1(\beta_i) = \alpha_i$ $(i = 1, 2, \ldots, p)$ (cf. Fig. 2.5). Then the ambient isotopy h_t assures that D' is isotoped so that D' is disjoint from $\bar{\alpha}$. □

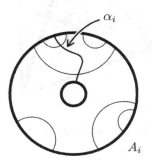

Fig. 2.5

In the remainder of these notes, let M be a compact connected orientable 3-manifold.

Definition 2.1.10. Let $(\partial_1 M, \partial_2 M)$ be a partition of the boundary components of M. A triplet $(C_1, C_2; S)$ is called a *Heegaard splitting* of $(M; \partial_1 M, \partial_2 M)$ if C_1 and C_2 are compression bodies with $C_1 \cup C_2 = M$, $\partial_- C_1 = \partial_1 M$, $\partial_- C_2 = \partial_2 M$ and $C_1 \cap C_2 = \partial_+ C_1 = \partial_+ C_2 = S$. Such a surface S is called a *Heegaard surface*. The *genus* of a Heegaard splitting is defined to be the genus of the Heegaard surface.

Theorem 2.1.11. *For any partition $(\partial_1 M, \partial_2 M)$ of the boundary components of M, there is a Heegaard splitting of $(M; \partial_1 M, \partial_2 M)$.*

Proof. It follows from Theorem 1.3.12 that M is triangulated, that is, there is a finite simplicial complex K which is homeomorphic to M. Let K' be a barycentric subdivision of K and K_1 the 1-skeleton of K. Here, a 1-*skeleton* of K is a union of the vertices and edges of K. Let $K_2 \subset K'$ be the dual 1-skeleton (see Fig. 2.6). Then each of K_i $(i = 1, 2)$ is a finite graph in M.

Case 1. $\partial M = \emptyset$.

Recall that K_1 consists of 0-simplices and 1-simplices. Set $C_1 = \eta(K_1; M)$ and $C_2 = \eta(K_2; M)$. Note that a regular neighborhood of a 0-simplex corresponds to a 0-handle and that a regular neighborhood of a

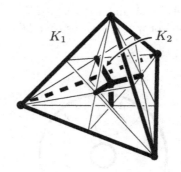

Fig. 2.6   A 1-skeleton K_1 of K and its dual 1-skeleton K_2.

1-simplex corresponds to a 1-handle. Hence C_1 is a handlebody. Similarly, we see that C_2 is also a handlebody. Then we see that $C_1 \cup C_2 = M$ and $C_1 \cap C_2 = \partial C_1 = \partial C_2$. Hence $(C_1, C_2; S)$ is a Heegaard splitting of M with $S = C_1 \cap C_2$.

Case 2. $\partial M \neq \emptyset$.

In this case we first take the barycentric subdivision of K and use the same notation K. Recall that K' is the barycentric subdivision of K. Note that no 3-simplices of K intersect both $\partial_1 M$ and $\partial_2 M$. Let $N(\partial_2 M)$ be a union of the 3-simplices in K' intersecting $\partial_2 M$. Then $N(\partial_2 M)$ is homeomorphic to $\partial_2 M \times [0, 1]$, where $\partial_2 M \times \{0\}$ is identified with $\partial_2 M$. Set $\partial_2' M = \partial_2 M \times \{1\}$. Let $\bar{K}_1$ ($\bar{K}_2$ resp.) be the maximal sub-complex of K_1 (K_2 resp.) such that $\bar{K}_1$ ($\bar{K}_2$ resp.) is disjoint from $\partial_2' M$ ($\partial_1 M$ resp.) (cf. Fig. 2.7).

Set $C_1 = \eta(\partial_1 M \cup \bar{K}_1; M)$. Then we can see that $C_1 = \eta(\partial_1 M; M) \cup \eta(\bar{K}_1; M)$. Note again that a regular neighborhood of a 0-simplex corresponds to a 0-handle and that a regular neighborhood of a 1-simplex corresponds to a 1-handle. Hence C_1 is obtained from $\partial_1 M \times [0, 1]$ by attaching 0-handles and 1-handles and therefore C_1 is a compression body with $\partial_- C_1 = \partial_1 M$. Set $C_2 = \eta(N(\partial_2 M) \cup \bar{K}_2; M)$. By the same argument, we can see that C_2 is a compression body with $\partial_- C_2 = \partial_2 M$. Since $C_1 \cup C_2 = M$ and $C_1 \cap C_2 = \partial C_1 = \partial C_2$, we see that $(C_1, C_2; S)$ is a Heegaard splitting of M with $S = C_1 \cap C_2$.   □

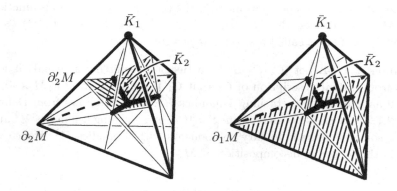

Fig. 2.7   The maximal sub-complex $\bar{K}_i$ of K_i such that $\bar{K}_i$ is disjoint from $\partial'_2 M$ if $i = 1$ and is disjoint from $\partial_1 M$ if $i = 2$, where a shaded portion in the figure is a part of $\partial'_2 M$ (left) or is a part of $\partial_1 M$ (right).

We now introduce alternative viewpoints on Heegaard splittings via the remarks below.

Definition 2.1.12. Let C be a compression body. A finite graph Σ in C is called a *spine* of C if $C \setminus (\partial_- C \cup \Sigma)$ is homeomorphic to $\partial_+ C \times [0, 1)$ and every vertex of valence one is in $\partial_- C$ (cf. Fig. 2.8).

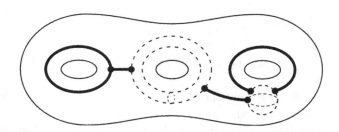

Fig. 2.8   A spine of a compression body.

Remark 2.1.13. For a Heegaard splitting $(C_1, C_2; S)$ of $(M; \partial_1 M, \partial_2 M)$, let Σ_i be a spine of C_i, and set $\Sigma'_i = \partial_i M \cup \Sigma_i$ ($i = 1, 2$). Then

$$M \setminus (\Sigma'_1 \cup \Sigma'_2) = (C_1 \setminus \Sigma'_1) \cup_S (C_2 \setminus \Sigma'_2) \cong S \times (0, 1),$$

where "$\cong$" means "homeomorphic to". Hence there is a continuous function $f : M \to [0,1]$ such that $f^{-1}(0) = \Sigma_1'$, $f^{-1}(1) = \Sigma_2'$ and $f^{-1}(t) \cong S$ ($0 < t < 1$). This is called a *sweep-out picture*.

Remark 2.1.14. Let $(C_1, C_2; S)$ be a Heegaard splitting as usual. It follows from a dual description of C_1 that C_1 is obtained from $\partial_1 M \times [0,1]$ and 0-handles $\mathcal{H}^0$ by attaching 1-handles $\mathcal{H}^1$. It also follows from Definition 2.1.2 that C_2 is obtained from $S \times [0,1]$ by attaching 2-handles $\mathcal{H}^2$ and filling some 2-sphere boundary components with 3-handles $\mathcal{H}^3$. Hence we obtain the following decomposition of M:

$$M = \partial_1 M \times [0,1] \cup \mathcal{H}^0 \cup \mathcal{H}^1 \cup S \times [0,1] \cup \mathcal{H}^2 \cup \mathcal{H}^3.$$

Collapsing $S \times [0,1]$ to S, we have

$$M = \partial_1 M \times [0,1] \cup \mathcal{H}^0 \cup \mathcal{H}^1 \cup_S \mathcal{H}^2 \cup \mathcal{H}^3.$$

This is called a *handle decomposition of M induced from* $(C_1, C_2; S)$.

2.2   Examples

2.2.1 $M = \mathbb{S}^3$

We consider the 3-sphere $\mathbb{S}^3$ as the unit sphere in $\mathbb{R}^4$, *i.e.*,

$$\mathbb{S}^3 = \{(x_1, x_2, x_3, x_4) \in \mathbb{R}^4 : x_1^2 + x_2^2 + x_3^2 + x_4^2 = 1\}.$$

We decompose $\mathbb{S}^3$ as follows: Let $h : \mathbb{R}^4 \to \mathbb{R}$ be projection onto the fourth coordinate. Then h sends $(x_1, x_2, x_3, x_4) \in \mathbb{S}^3$ to $x_4 \in [-1,1]$. Clearly $\mathbb{S}^3 \cap h^{-1}(-1)$ is the single point $(0,0,0,-1)$ which we call the *south pole* of $\mathbb{S}^3$. If $t \in (-1,1)$, then $\mathbb{S}^3 \cap h^{-1}(t)$ is the set of points

$$\{(x_1, x_2, x_3, 0) \in \mathbb{R}^4 : x_1^2 + x_2^2 + x_3^2 = 1 - t^2\},$$

i.e., a 3-ball of radius $\sqrt{1 - t^2}$. Note that its radius increases monotonically to one if $t \in (-1, 0]$, and it decreases monotonically if $t \in [0, 1)$. Similarly $\mathbb{S}^3 \cap h^{-1}(1)$ is the single point $(0, 0, 0, 1)$ which we call the *north pole* of $\mathbb{S}^3$. We will see later that this describes a sweep-out picture by proving each of the following is homeomorphic to a 3-ball (Exercise 2.2.1).

$$\mathbb{S}^3_+ = \{(x_1, x_2, x_3, x_4) \in \mathbb{S}^3 : x_4 \geq 0\},$$
$$\mathbb{S}^3_- = \{(x_1, x_2, x_3, x_4) \in \mathbb{S}^3 : x_4 \leq 0\}.$$

Hence we see that $(\mathbb{S}^3_+, \mathbb{S}^3_- ; S)$ is a genus zero Heegaard splitting of $\mathbb{S}^3$, where $S = \mathbb{S}^3 \cap h^{-1}(0)$ and hence $S = \partial \mathbb{S}^3_+ = \partial \mathbb{S}^3_-$.

Exercise 2.2.1. Show that $\mathbb{S}^3_\pm$ is homeomorphic to a 3-ball.

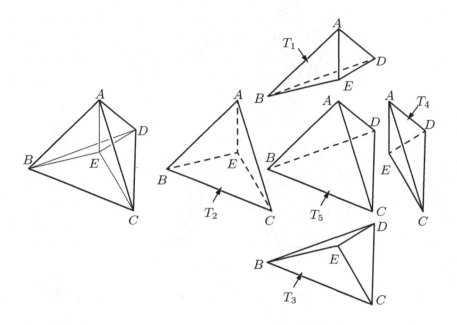

Fig. 2.9 $\mathbb{S}^3$ can be considered as the boundary of a 4-simplex.

We give another approach to obtaining a genus zero Heegaard splitting of $\mathbb{S}^3$: Recall that $\mathbb{S}^3$ can be considered as the boundary of a 4-ball. The 4-ball is homeomorphic to a 4-simplex, *i.e.*, a 5-cell in $\mathbb{R}^4$ with corners

$$A = (0,0,0,0), B = (0,0,0,1), C = (0,0,1,0), D = (0,1,0,0), E = (1,0,0,0).$$

The boundary of the 5-cell consists of five tetrahedra, say $T_1, \ldots, T_5$ as illustrated in Fig. 2.9. Then, for example, $T_1 \cup T_2 \cup T_3$ and $T_4 \cup T_5$ are

3-balls which share their boundaries. See Fig. 2.10.

Remark 2.2.2. It follows from [Alexander (1924)] that a 3-manifold obtained from two 3-balls by identifying their boundaries is always homeomorphic to $\mathbb{S}^3$.

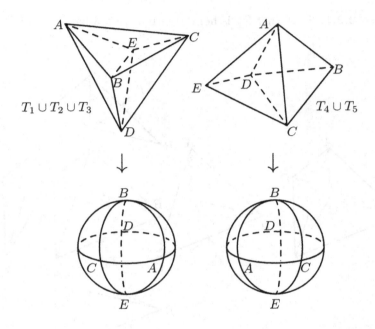

Fig. 2.10   Another approach to obtain a genus zero Heegaard splitting of $\mathbb{S}^3$.

2.2.2 $M = \mathbb{S}^2 \times \mathbb{S}^1$

Note that $\mathbb{S}^2 \times \mathbb{S}^1$ is obtained from $\mathbb{S}^2 \times [0, 1]$ by identifying $\{p\} \times \{0\}$ with $\{p\} \times \{1\}$ for each point $p \in \mathbb{S}^2$. Then a regular neighborhood of $\{p\} \times \mathbb{S}^1$ in $\mathbb{S}^2 \times \mathbb{S}^1$ is homeomorphic to a solid torus. We also see that its exterior is also homeomorphic to a solid torus (cf. Fig. 2.11). Hence these solid tori give a genus one Heegaard splitting of $\mathbb{S}^2 \times \mathbb{S}^1$.

Remark 2.2.3. For each $t \in [0, 1]$, a 2-sphere $\mathbb{S}^2 \times \{t\}$ in $\mathbb{S}^2 \times \mathbb{S}^1$ intersects each solid torus in a single meridian disk of the solid torus. Equivalently,

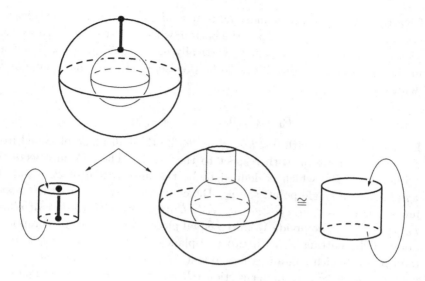

Fig. 2.11   A genus one Heegaard splitting of $\mathbb{S}^2 \times \mathbb{S}^1$.

each solid torus contains a meridian disk such that those two disks share their boundaries.

2.2.3 *Lens spaces*

We consider $\mathbb{B}^3$ as the unit 3-ball in $\mathbb{R}^3$, *i.e.*,

$$\mathbb{B}^3 = \{(x, y, z) \in \mathbb{R}^3 : x^2 + y^2 + z^2 \leq 1\}.$$

As usual, $\mathbb{S}^2_+$ ($\mathbb{S}^2_-$, respectively) denotes the upper (lower, respectively) hemisphere of $\partial \mathbb{B}^3$ consisting of the points with $z \geq 0$ ($z \leq 0$, respectively). Let (p, q) be a pair of coprime integers with $0 \leq q < p$. For a point $v \in \mathbb{S}^2_+$, $r_{p,q}(v)$ is defined to be the point in $\mathbb{S}^2_-$ obtained by a $\frac{2\pi q}{p}$-rotation with respect to the z-axis and then by the reflection with respect to the xy-plane (cf. the upper left of Fig. 2.12). Let $M_{p,q}$ be a 3-manifold obtained from $\mathbb{B}^3$ by identifying v with $r_{p,q}(v)$ for each point $v \in \mathbb{S}^2_+$. We verify that $M_{p,q}$ admits a genus one Heegaard splitting. Set

$$V_1 = \{(x, y, z) \in \mathbb{B}^3 : x^2 + y^2 \leq 1/4\},$$
$$V_2 = \{(x, y, z) \in \mathbb{B}^3 : x^2 + y^2 \geq 1/4\}.$$

Clearly V_1 becomes homeomorphic to a solid torus after the identification above, *i.e.*, attaching the top to the bottom with a $\frac{2\pi q}{p}$-rotation. To see that V_2 is also a solid torus, we cut V_2 vertically into p equal pieces as described in the upper right of Fig. 2.12. To be precise, let P_0 be a half-plane in $\mathbb{R}^3$ defined as

$$P_0 = \{(x, 0, z) \in \mathbb{R}^3 : x \geq 0\}.$$

For each integer k with $0 < k < p$, let P_k be the half-plane obtained from P_0 by a $\frac{2\pi k}{p}$-rotation with respect to the z-axis. Then P_k intersects the "equator" of $\mathbb{B}^3$, which is defined to be the intersection of $\partial \mathbb{B}^3$ and the xy-plane, in a single point, say v_k. Hence $v_0 \cup \cdots \cup v_{p-1}$ cuts the equator into p segments $v_0 v_1, v_1 v_2, \ldots, v_{p-1} v_0$. Cutting V_2 with these half planes $P_0 \cup \cdots \cup P_{p-1}$, we divide V_2 into p equal pieces. Let B_k be the component obtained by cutting V_2 with two half-planes $P_k \cup P_{k+1}$, where k is taken mod p, and which contains the segment $v_k v_{k+1}$. Since we identify $v \in \mathbb{S}^2_+$ with $r_{p,q}(v) \in \mathbb{S}^2_-$, an upper section $\partial B_k \cap \mathbb{S}^2_+$, say F_k, is identified with a lower section $\partial B_{k+q} \cap \mathbb{S}^2_-$. See the middle of Fig. 2.12 which describes the case of $(p, q) = (7, 3)$. We now do these identifications. The lower left of Fig. 2.12 describes $B_0 \cup B_3$ attached along F_0. We next attach B_4 to $B_0 \cup B_3$ along F_3. Attaching all pieces similarly, we have the lower right of Fig. 2.12 which becomes a solid torus after performing the remaining identifications. Hence V_1 and V_2 give a genus one Heegaard splitting of $M_{p,q}$. The manifold $M_{p,q}$ obtained as above is called the *lens space of type* (p, q).

Remark 2.2.4. Let V_i ($i = 1, 2$) be solid tori above and D_i meridian disks of V_i. Then the geometric intersection number of ∂D_1 and ∂D_2 is equal to p.

Exercise 2.2.5. Verify Remark 2.2.4.

Exercise 2.2.6. Show that the lens space of type $(1, 0)$ is homeomorphic to $\mathbb{S}^3$.

Remark 2.2.7. Let $M_{p,q}$ and $M_{p',q'}$ be lens spaces of type (p, q) and (p', q') respectively. It is shown in [Reidemeister (1935)] that $M_{p,q}$ and $M_{p',q'}$ are (PL) homeomorphic if and only if $p = p'$ and $qq' \equiv \pm 1 \pmod{p}$. See also [Brody (1960)].

Remark 2.2.8. The lens space of type $(2, 1)$ is also called the *real projective space*.

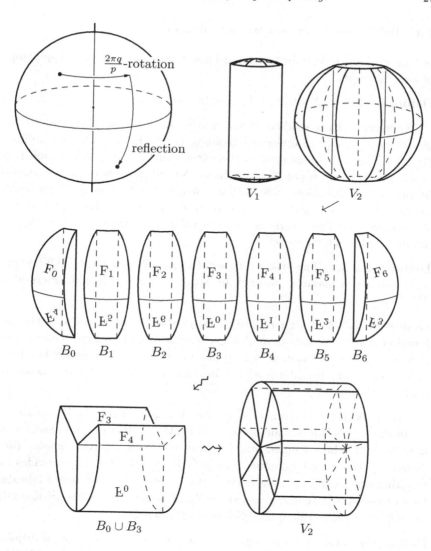

Fig. 2.12   A genus one Heegaard splitting of $M_{p,q}$ with $(p, q) = (7, 3)$.

2.3   Bridge position and tunnel number

In this section we describe a standard construction for Heegaard splittings of knot exteriors.

Definition 2.3.1. A *knot* is a PL 1-submanifold in $\mathbb{S}^3$.

Knots are widely studied in a variety of subjects ranging from low dimensional topology to molecular biology. See [Adams (1994)] or [Rolfsen (1990)]. The techniques used in their study include algebraic, graph theoretical and purely topological methods. Visualizing knots proves fruitful in guiding our intuition. Recall that a *height function* on $\mathbb{S}^3$ is a smooth function $h : \mathbb{S}^3 \to \mathbb{R}$ with exactly two critical points, a minimum $-\infty$ and a maximum ∞. The inverse image of a value other than the maximum or minimum of h is a 2-sphere.

Definition 2.3.2. The *bridge number* of a knot K, denoted by $b(K)$, is the least number of maxima required for $h|_K$, where h is a height function on $\mathbb{S}^3$.

Definition 2.3.3. A knot K is in *bridge position* with respect to the height function h if every maximum of $h|_K$ occurs above every minimum of $h|_K$. If K is in bridge position with respect to h, then a *bridge sphere* is a level surface of h that lies above all the minima and below all the maxima of $h|_K$.

We think of the maxima of K as the "bridges" of K. See Fig. 2.13.

In the 1950's Horst Schubert studied bridge number of knots and found it to be well-behaved under standard constructions involving knots. (See [Schubert (1954)] and [Schultens (2003)].) Bridge number also provides interesting classes of 3-manifolds. For instance, even the exteriors of 2-bridge knots constitute a rich and varied family of 3-manifolds. (See [Kobayashi (2001)], [Schubert (1956)] and [Schultens (2001/02)].)

Definition 2.3.4. A *tunnel system* for a knot K is a collection of disjoint arcs $\mathcal{T} = t_1 \cup \ldots \cup t_n$ properly embedded in $E(K) = \mathrm{cl}(\mathbb{S}^3 \setminus \eta(K, \mathbb{S}^3))$ such that $\mathrm{cl}(E(K) \setminus \eta(\mathcal{T}; E(K)))$ is a handlebody. The *tunnel number of* K, denoted by $t(K)$, is the least number of arcs required in a tunnel system for K.

For 2-bridge knots, there is a tunnel system consisting of a single arc. Fig. 2.14 allows us to visualize the complement of a 2-bridge knot K together with one arc, say τ. The exterior of a regular neighborhood of $K \cup \tau$

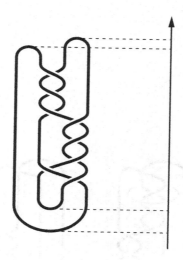

Fig. 2.13   A knot in a 2-bridge position.

in $\mathbb{S}^3$ is unchanged, the final figure illustrates that this exterior is a genus two handlebody. This also implies that $\eta(K \cup \tau, \mathbb{S}^3)$ and its exterior give a genus two Heegaard splitting of $\mathbb{S}^3$.

More generally, an n-bridge knot will have a tunnel system consisting of $n-1$ arcs which will give genus n Heegaard splittings both of its complement and of $\mathbb{S}^3$. Tunnel number is not well-behaved under standard constructions involving knots. (See [Kobayashi (1994)], [Morimoto (2000)], [Morimoto (1995b)].) However, some of the pathologies are understood. (See [Kobayashi and Saito (2010)], [Morimoto (1995a)] and [Morimoto and Schultens (2000)].)

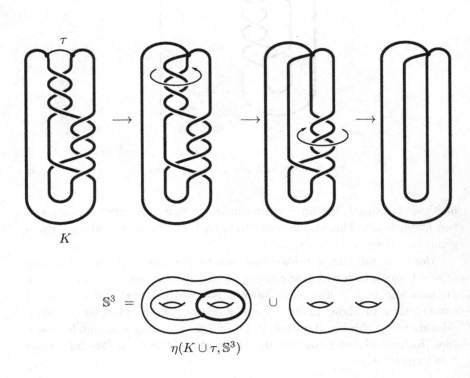

Fig. 2.14 The exterior of a regular neighborhood of $K \cup \tau$ in $\mathbb{S}^3$ is homeomorphic to a genus two handlebody.

Chapter 3

Properties of Heegaard splittings

In the 1930's, Reidemeister and Singer independently proved theorems concerning a construction, see below, called stabilization. (See [Reidemeister (1935)] and [Singer (1933)].) Their motivation came from the lack of uniqueness of Heegaard splittings of 3-manifolds. In the 1960's, Haken and Waldhausen continued investigating the questions asked by Reidemeister and Singer. (See [Haken (1968)] and [Waldhausen (1968)].) In the late 1980's Casson and Gordon (see [Casson and Gordon (1987)]) further continued this study by introducing the notion of strong irreducibility and proving an extension of a key lemma due to Haken. In this chapter we discuss the concepts central to the work of Reidemeister, Singer, Haken, Waldhausen, Casson and Gordon. In the next chapter we will examine two well known theorems more closely.

3.1 Reducibility, weak reducibility and stabilization

Definition 3.1.1. Let M be a connected compact 3-manifold possibly with boundary and $(C_1, C_2; S)$ a Heegaard splitting of $(M; \partial_1 M, \partial_2 M)$.

(1) The Heegaard splitting $(C_1, C_2; S)$ is said to be *reducible* if there are meridian disks D_i $(i = 1, 2)$ of C_i with $\partial D_1 = \partial D_2$. The Heegaard splitting $(C_1, C_2; S)$ is said to be *irreducible* if $(C_1, C_2; S)$ is not reducible.

(2) The Heegaard splitting $(C_1, C_2; S)$ is said to be *weakly reducible* if there are meridian disks D_i $(i = 1, 2)$ of C_i with $\partial D_1 \cap \partial D_2 = \emptyset$. The Heegaard splitting $(C_1, C_2; S)$ is said to be *strongly irreducible* if $(C_1, C_2; S)$ is not weakly reducible.

(3) The Heegaard splitting $(C_1, C_2; S)$ is said to be *∂-reducible* if there is a disk D properly embedded in M such that $D \cap S$ is an essential

simple closed curve in S. Such a disk D is called a ∂-*reducing disk* for $(C_1, C_2; S)$.

(4) The Heegaard splitting $(C_1, C_2; S)$ is said to be *stabilized* if there are meridian disks D_i $(i = 1, 2)$ of C_i such that ∂D_1 and ∂D_2 intersect transversely in a single point. Such a pair of disks is called a *cancelling pair of disks* for $(C_1, C_2; S)$.

Example 3.1.2.

(1) The genus zero Heegaard splitting of $\mathbb{S}^3$ described in Section 2.2.1 is strongly irreducible.
(2) The genus one Heegaard splitting of $\mathbb{S}^2 \times \mathbb{S}^1$ described in Section 2.2.2 is reducible.
(3) The genus one Heegaard splitting of the lens space of type $(1, 0)$ described in Section 2.2.2 is stabilized and is irreducible.
(4) Any genus one Heegaard splitting of the lens space of type (p, q) described in Section 2.2.2 is strongly irreducible unless $(p, q) = (1, 0)$.

Example 3.1.3. Let $(C_1, C_2; S)$ be a Heegaard splitting such that S is a 2-sphere and that each of $\partial_- C_i$ $(i = 1, 2)$ consists of two 2-spheres. Note that there does not exist an essential disk in C_i. Hence $(C_1, C_2; S)$ is strongly irreducible.

The following is an easy observation directly obtained from the definitions above.

Observation 3.1.4. Every reducible Heegaard splitting is weakly reducible.

Though the converse of Observation 3.1.4 is not always true, we can say:

Remark 3.1.5. Let $(C_1, C_2; S)$ be a weakly reducible Heegaard splitting. If the Heegaard surface S is of genus less than three, then $(C_1, C_2; S)$ is reducible.

Exercise 3.1.6. Show Remark 3.1.5.

Suppose that $(C_1, C_2; S)$ is stabilized, and let D_i $(i = 1, 2)$ be disks as in (4) of Definition 3.1.1. Note that since ∂D_1 intersects ∂D_2 transversely in a single point, we see that each of ∂D_i $(i = 1, 2)$ is non-separating in S and hence each of D_i $(i = 1, 2)$ is non-separating in C_i. Set $C_1' =$

$cl(C_1 \setminus \eta(D_1; C_1))$ and $C_2' = C_2 \cup \eta(D_1; C_1)$. Then each of C_i' $(i = 1, 2)$ is a compression body with $\partial_+ C_1' = \partial_+ C_2'$ (cf. (6) of Remark 2.1.4). Set $S' = \partial_+ C_1' (= \partial_+ C_2')$. Then we obtain a Heegaard splitting $(C_1', C_2'; S')$ of M with genus$(S') =$ genus$(S) - 1$. Conversely, $(C_1, C_2; S)$ is obtained from $(C_1', C_2'; S')$ by adding a trivial handle. We say that $(C_1, C_2; S)$ is obtained from $(C_1', C_2'; S')$ by *stabilization*. Note that genus$(S) =$ genus$(S') + 1$.

Remark 3.1.7. A genus one Heegaard splitting of the lens space of type $(1, 0)$, which is homeomorphic to $\mathbb{S}^3$, described in Section 2.2.2 is obtained from the genus zero Heegaard splitting described in Section 2.2.1 by stabilization.

Exercise 3.1.8. Verify Remark 3.1.7.

Lemma 3.1.9. *Let $(C_1, C_2; S)$ be a Heegaard splitting of $(M; \partial_1 M, \partial_2 M)$ such that the genus of S is at least two. If $(C_1, C_2; S)$ is stabilized, then $(C_1, C_2; S)$ is reducible.*

Proof. Suppose that $(C_1, C_2; S)$ is stabilized, and let D_i $(i = 1, 2)$ be meridian disks of C_i such that ∂D_1 intersects ∂D_2 transversely in a single point. Then $\partial \eta(\partial D_1 \cup \partial D_2; S)$ bounds a disk D_i' in C_i for each $i = 1$ and 2. In fact, for each $(i, j) = (1, 2)$ and $(2, 1)$, the disk D_i' is obtained from two parallel copies of D_i by connecting them with a "band" along ∂D_j (cf. Fig. 3.1).

Note that $\partial D_1' = \partial D_2'$ cuts S into a torus with a single hole and the other surface S'. Since the genus of S is at least two, we see that the genus of S' is at least one. Hence $\partial D_1' = \partial D_2'$ is essential in S and therefore $(C_1, C_2; S)$ is reducible. $\qquad\qquad\square$

Definition 3.1.10. Let $(C_1, C_2; S)$ be a Heegaard splitting of $(M; \partial_1 M, \partial_2 M)$.

(1) Suppose that M is homeomorphic to $\mathbb{S}^3$. We call $(C_1, C_2; S)$ a *trivial splitting* if both C_1 and C_2 are 3-balls.
(2) Suppose that M is not homeomorphic to $\mathbb{S}^3$. We call $(C_1, C_2; S)$ a *trivial splitting* if C_i is a trivial compression body for $i = 1$ or 2.

Remark 3.1.11. Suppose that M is not homeomorphic to $\mathbb{S}^3$. If $(M; \partial_1 M, \partial_2 M)$ admits a trivial splitting $(C_1, C_2; S)$, then it is easy to see that M is a compression body. Particularly, if C_2 (C_1 resp.) is trivial, then $\partial_- M = \partial_1 M$ and $\partial_+ M = \partial_2 M$ ($\partial_- M = \partial_2 M$ and $\partial_+ M = \partial_1 M$ resp.).

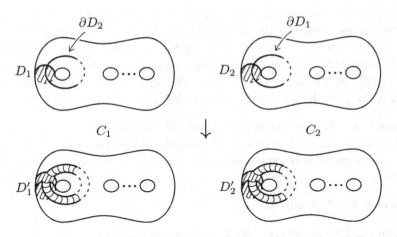

Fig. 3.1 The disk D_i' is obtained from two parallel copies of D_i by connecting them with a "band" along ∂D_j for each $(i,j) = (1,2)$ and $(2,1)$.

Lemma 3.1.12. *Let $(C_1, C_2; S)$ be a non-trivial Heegaard splitting of $(M; \partial_1 M, \partial_2 M)$. If $(C_1, C_2; S)$ is ∂-reducible, then $(C_1, C_2; S)$ is weakly reducible.*

Proof. Let D be a ∂-reducing disk for $(C_1, C_2; S)$. (Hence $D \cap S$ is an essential simple closed curve in S.) Set $D_1 = D \cap C_1$ and $A_2 = D \cap C_2$. Exchanging subscripts, if necessary, we may suppose that D_1 is a meridian disk of C_1 and A_2 is a spanning annulus in C_2. Note that $A_2 \cap \partial_- C_2$ is an essential simple closed curve in the component of $\partial_- C_2$ containing $A_2 \cap \partial_- C_2$. Since C_2 is non-trivial, there is a meridian disk of C_2. It follows from Lemma 2.1.6 that we can choose a meridian disk D_2 of C_2 with $D_2 \cap A_2 = \emptyset$. This implies that $D_1 \cap D_2 = \emptyset$. Hence $(C_1, C_2; S)$ is weakly reducible. $\qquad \square$

3.2 Haken's Lemma

In this section we prove the following.

Theorem 3.2.1. *Let $(C_1, C_2; S)$ be a Heegaard splitting of $(M; \partial_1 M, \partial_2 M)$.*

(1) If M is reducible, then $(C_1, C_2; S)$ is reducible or C_i is reducible for $i = 1$ or 2.

(2) If M is ∂-reducible, then $(C_1, C_2; S)$ is ∂-reducible.

Note that the first statement of Theorem 3.2.1 is called Haken's Lemma and proved by Haken in [Haken (1968)]. The second statement of Theorem 3.2.1 is proved by Casson and Gordon in [Casson and Gordon (1987)]. We note that the argument in this section is based on [Scharlemann and Thompson (1994a)].

We first prove the following proposition, whose statement is weaker than that of Theorem 3.2.1, after showing some lemmas.

Proposition 3.2.2. *If M is reducible or ∂-reducible, then $(C_1, C_2; S)$ is reducible, ∂-reducible, or C_i is reducible for $i = 1$ or 2.*

We give a proof of Proposition 3.2.2 by using Otal's idea (cf. [Otal (1991)]) of viewing the Heegaard splittings as a graph in the three dimensional space.

Edge slides of graphs. Let Γ be a finite graph in a 3-manifold M. Choose an edge σ of Γ and let $\bar{\Gamma}$ be a subgraph of Γ obtained by removing the edge σ from Γ. We suppose that σ joins $\bar{\Gamma}$ to itself. Let p_1 and p_2 be the vertices of Γ incident to σ. We may suppose that $\sigma \cap \partial\eta(\bar{\Gamma}; M)$ consists of two points, say $\bar{p}_1$ and $\bar{p}_2$ (cf. Fig. 3.2). Then $\bar{p}_1 \cup \bar{p}_2$ divides σ into three sub-edges, say α_0, α_1 and α_2 with $\partial\alpha_0 = \bar{p}_1 \cup \bar{p}_2$, $\partial\alpha_1 = p_1 \cup \bar{p}_1$ and $\partial\alpha_2 = p_2 \cup \bar{p}_2$.

Fig. 3.2   A pair of two points $\bar{p}_1 \cup \bar{p}_2$ divides σ into three sub-edges.

Take a path γ on $\partial\eta(\bar{\Gamma}; M)$ with $\partial\gamma \ni \bar{p}_1$. Let $\bar{\sigma}$ be an edge obtained from $\gamma \cup \alpha_0 \cup \alpha_2$ by adding a "straight short arc" in $\eta(\bar{\Gamma}; M)$ connecting the endpoint of γ other than $\bar{p}_1$ and a point p_1' in the interior of an edge of $\bar{\Gamma}$ (cf. Fig. 3.3). Let Γ' be a graph obtained from $\bar{\Gamma} \cup \bar{\sigma}$ by adding p_1' as a vertex. Then we say that Γ' is obtained from Γ by an *edge slide* on σ.

If p_1 is a trivalent vertex, then it is natural for us not to regard p_1 as a vertex of Γ'. Particularly, the deformation of Γ which is depicted as in Fig. 3.3 is realized by an edge slide and an isotopy. This deformation is called a *Whitehead move*.

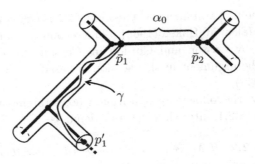

Fig. 3.3   An edge $\bar{\sigma}$ is obtained from $\gamma \cup \alpha_0 \cup \alpha_2$ by adding a "straight short arc".

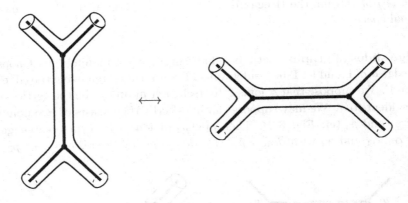

Fig. 3.4   A Whitehead move.

We now prepare some lemmas to prove Proposition 3.2.2. Let Σ be a spine of C_1. Note that $\eta(\partial_- C_1 \cup \Sigma; M)$ is obtained from regular neighborhoods of $\partial_- C_1$ and the vertices of Σ by attaching 1-handles corresponding to the edges of Σ. Set $\Sigma_\eta = \eta(\Sigma; M)$. The notation h_v^0, called a *vertex* of Σ_η, means a regular neighborhood of a vertex v of Σ. Also, the notation h_σ^1, called an *edge* of Σ_η, means a 1-handle corresponding to an edge σ of Σ. Let $\Delta = D_1 \cup \cdots \cup D_k$ be a minimal complete meridian system of C_2.

Let P be a reducing 2-sphere or a ∂-reducing disk of M. If P is a ∂-reducing disk, we may assume that $\partial P \subset \partial_- C_2$ by changing subscripts. We may assume that P intersects Σ and Δ transversely. Set $\Gamma = P \cap (\Sigma_\eta \cup \Delta)$. We note that Γ is a union of disks $P \cap \Sigma_\eta$ and a union of arcs and simple closed curves $P \cap \Delta$ in P. We choose P, Σ and Δ so that the pair $(|P \cap$

$\Sigma|, |P \cap \Delta|$) is minimal with respect to lexicographic order.

Lemma 3.2.3. *Each component of $P \cap \Delta$ is an arc.*

Proof. For some disk component, say D_1, of Δ, suppose that $P \cap D_1$ contains a simple closed curve. Let α be such a component of $P \cap D_1$ which is innermost in D_1, and let δ_α be an innermost disk for α. Let δ'_α be a disk in P with $\partial \delta'_\alpha = \alpha$. Set $P' = (P \setminus \delta'_\alpha) \cup \delta_\alpha$ if P is a ∂-reducing disk, or set $P' = (P \setminus \delta'_\alpha) \cup \delta_\alpha$ and $P'' = \delta'_\alpha \cup \delta_\alpha$ if P is a reducing 2-sphere. If P is a ∂-reducing disk, then P' is also a ∂-reducing disk. If P is a reducing 2-sphere, then either P' or P'', say P', is a reducing 2-sphere. Moreover, we can isotope P' so that $(|P' \cap \Sigma|, |P' \cap \Delta|) < (|P \cap \Sigma|, |P \cap \Delta|)$. This contradicts the minimality of $(|P \cap \Sigma|, |P \cap \Delta|)$. $\qquad \square$

It follows from Lemma 3.2.3 that Γ is regarded as a graph in P which consists of *fat-vertices* $P \cap \Sigma_\eta$ and edges $P \cap \Delta$. An edge of the graph Γ is called a *loop* if the edge joins a fat-vertex of Γ to itself, and a loop is said to be *inessential* if the loop cuts off a disk from $\mathrm{cl}(P \setminus \Sigma_\eta)$ whose interior is disjoint from $\Gamma \cap \Sigma_\eta$.

Lemma 3.2.4. Γ *does not contain an inessential loop.*

Proof. Suppose that Γ contains an inessential loop μ. Then μ cuts off a disk δ_μ from $\mathrm{cl}(P \setminus \Sigma_\eta)$ such that the interior of δ_μ is disjoint from $\Gamma \cap \Sigma_\eta$ (cf. Fig. 3.5).

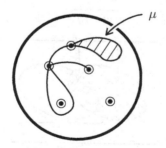

Fig. 3.5 An inessential loop μ.

We may assume that $\delta_\mu \cap \Delta = \delta_\mu \cap D_1$. Then μ cuts D_1 into two disks D'_1 and D''_1 (cf. Fig. 3.6).

Let C'_2 be the component, which is obtained by cutting C_2 along Δ, such that C'_2 contains δ_μ. Let D_1^+ be the copy of D_1 in C'_2 with $D_1^+ \cap \delta_\mu \neq \emptyset$

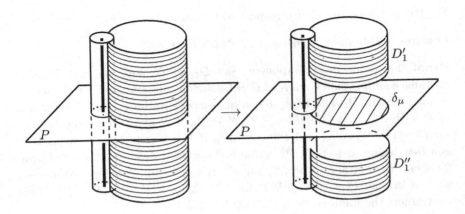

Fig. 3.6   An inessential loop μ cuts D_1 into two disks D_1' and D_1''.

and D_1^- the other copy of D_1. Note that C_2' is a 3-ball or (a component of $\partial_- C_2) \times [0,1]$. This shows that there is a disk δ_μ' in $\partial_+ C_2'$ such that $\partial \delta_\mu = \partial \delta_\mu'$ and that $\partial \delta_\mu \cup \partial \delta_\mu'$ bounds a 3-ball in C_2'. Note that $\delta_\mu' \cap D_1^+ \neq \emptyset$. By changing superscripts, if necessary, we may assume that $\delta_\mu' \supset D_1'$ (cf. Fig. 3.7).

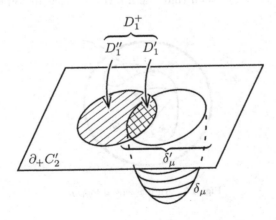

Fig. 3.7   The disk δ_μ' is assumed to contain D_1'.

Set $D_0 = \delta_\mu \cup D_1'$ if $\delta_\mu' \cap D_1^- \neq \emptyset$, and $D_0 = \delta_\mu \cup D_1''$ if $\delta_\mu' \cap D_1^- = \emptyset$. We may regard D_0 as a disk properly embedded in C_2. Set $\Delta' = D_0 \cup D_2 \cup \cdots \cup D_k$. Then we see that Δ' is a minimal complete meridian system of C_2. We can further isotope D_0 slightly so that $|P \cap \Delta'| < |P \cap \Delta|$. This contradicts the minimality of $(|P \cap \Sigma|, |P \cap \Delta|)$. $\qquad \square$

A fat-vertex of Γ is said to be *isolated* if there are no edges of Γ incident to the fat-vertex (cf. Fig. 3.8).

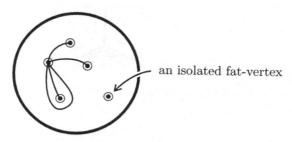
an isolated fat-vertex

Fig. 3.8   An example of an isolated fat-vertex.

Lemma 3.2.5. *If Γ has an isolated fat-vertex, then $(C_1, C_2; S)$ is reducible or ∂-reducible.*

Proof. Suppose that there is an isolated fat-vertex D_v of Γ. Recall that D_v is a component of $P \cap \Sigma_\eta$ which is a meridian disk of C_1. Note that D_v is disjoint from Δ (cf. Fig. 3.9).

Let C_2' be a component obtained by cutting C_2 along Δ such that $\partial C_2'$ contains ∂D_v. If ∂D_v bounds a disk D_v' in C_2', then D_v and D_v' show the reducibility of $(C_1, C_2; S)$. Otherwise, C_2' is (a closed orientable surface) $\times$ $[0, 1]$, and ∂D_v is a boundary component of a spanning annulus in C_2' and hence in C_2. Hence we see that $(C_1, C_2; S)$ is ∂-reducible. $\qquad \square$

Lemma 3.2.6. *Suppose that no fat-vertices of Γ are isolated. Then each fat-vertex of Γ is a base of a loop.*

Proof. Suppose that there is a fat-vertex D_w of Γ which is not a base of a loop. Since no fat-vertices of Γ are isolated, there is an edge of Γ incident to D_w. Let σ be the edge of Σ with $h_\sigma^1 \supset D_w$. (Recall that h_σ^1 is a 1-handle of Σ_η corresponding to σ.) Let D be a component of Δ with $\partial D \cap h_\sigma^1 \neq \emptyset$. Let C_w be a union of the arc components of $D \cap P$ which are incident to D_w. Let γ be an arc component of C_w which is outermost

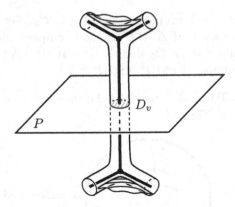

Fig. 3.9　An isolated fat-vertex D_v of Γ is regarded as a meridian disk of C_1 which is disjoint from Δ.

among the components of C_w. We call such an arc γ an *outermost edge for D_w of* Γ. Let $\delta_\gamma \subset D$ be a disk obtained by cutting D along γ whose interior is disjoint from the edges incident to D_w. We call such a disk δ_γ an *outermost disk for* (D_w, γ). (Note that δ_γ may intersect P transversely (cf. Fig. 3.10).) Let $D_{w'}(\neq D_w)$ be the fat-vertex of Γ attached to γ. Then we have the following three cases.

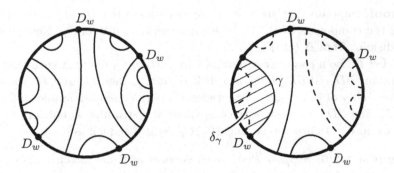

Fig. 3.10　An outermost edge γ for D_w and an outermost disk δ_γ for (D_w, γ).

Case 1.　$(\partial \delta_\gamma \setminus \gamma) \subseteq (h^1_\sigma \cap D)$.

In this case we can isotope σ along δ_γ to reduce $|P \cap \Sigma|$ (cf. Fig. 3.11).

Case 2.　$(\partial \delta_\gamma \setminus \gamma) \nsubseteq (h^1_\sigma \cap D)$ and $D_{w'} \not\subset (h^1_\sigma \cap D)$.

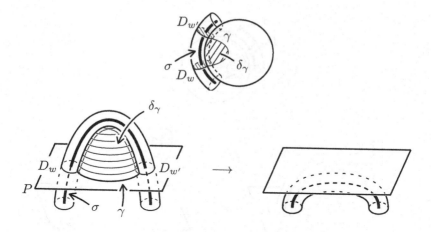

Fig. 3.11 Case 1. The edge σ is isotoped along δ_γ so that $|P \cap \Sigma|$ is reduced.

Let p be the vertex of Σ such that $p \cap \sigma \neq \emptyset$ and $h_p^0 \cap \delta_\gamma \neq \emptyset$. Let β be the component of $\text{cl}(\sigma \setminus D_w)$ which satisfies $\beta \cap p \neq \emptyset$. Then we can slide β along δ_γ so that β contains γ (cf. Fig. 3.12). We can further isotope β slightly to reduce $|P \cap \Sigma|$, a contradiction.

Case 3. $(\partial \delta_\gamma \setminus \gamma) \not\subseteq (h_\sigma^1 \cap D)$ and $D_{w'} \subset (h_\sigma^1 \cap D)$.

Let p and p' be the endpoints of σ. Let β and β' be the components of $\text{cl}(\sigma \setminus (D_w \cup D_{w'}))$ which satisfy $p \cap \beta \neq \emptyset$ and $p' \cap \beta' \neq \emptyset$. Suppose first that $p \neq p'$. Then we can slide β along δ_γ so that β contains γ (cf. Fig. 3.13). We can further isotope β slightly to reduce $|P \cap \Sigma|$, a contradiction.

Suppose next that $p = p'$. In this case we perform the following operation which is called a *broken edge slide*.

We first add $w' = D_{w'} \cap \Sigma$ as a vertex of Σ. Then w' cuts σ into two edges β' and $\text{cl}(\sigma \setminus \beta')$. Since γ is an outermost edge for D_w of Γ, we see that $\beta' \subset \beta$ (cf. Fig. 3.14). Hence we can slide $\text{cl}(\beta \setminus \beta')$ along δ_γ so that $\text{cl}(\beta \setminus \beta')$ contains γ. We now remove the vertex w' of Σ, that is, we regard a union of β' and $\text{cl}(\sigma \setminus \beta')$ as an edge of Σ again. Then we can isotope $\text{cl}(\sigma \setminus \beta')$ slightly to reduce $|P \cap \Sigma|$, a contradiction (cf. Fig. 3.15). $\square$

Proof of Proposition 3.2.2. It follows from Lemma 3.2.5 that if there is an isolated fat-vertex of Γ, then we have the conclusion of Proposition 3.2.2. Hence we suppose that no fat-vertices of Γ are isolated. Then it

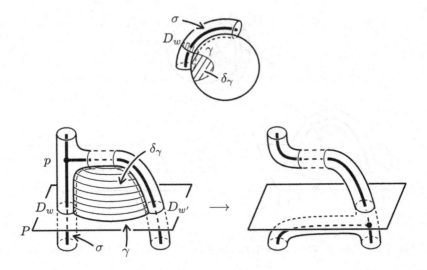

Fig. 3.12 Case 2. the component β is isotoped along δ_γ so that $|P \cap \Sigma|$ is reduced.

follows from Lemma 3.2.6 that each fat-vertex of Γ is a base of a loop. Let μ be a loop which is innermost in P. Then μ cuts a disk δ_μ from $\mathrm{cl}(P \setminus \Sigma_\eta)$. Since μ is essential (cf. Lemma 3.2.4), we see that δ_μ contains a fat-vertex of Γ. However, since μ is innermost, such a fat-vertex is not a base of any loop. Hence such a fat-vertex is isolated, a contradiction. This completes the proof of Proposition 3.2.2. □

Proof of (1) in Theorem 3.2.1. Suppose that M is reducible. Then it follows from Proposition 3.2.2 that $(C_1, C_2; S)$ is reducible or ∂-reducible, or C_i is reducible for $i = 1$ or 2. If $(C_1, C_2; S)$ is reducible or C_i is reducible for $i = 1$ or 2, then we are done. Hence we may assume that both C_1 and C_2 are irreducible and that $(C_1, C_2; S)$ is ∂-reducible. By induction on the genus of a Heegaard surface S, we prove that $(C_1, C_2; S)$ is reducible.

Suppose that $\mathrm{genus}(S) = 0$. Since C_i ($i = 1, 2$) are irreducible, we see that each of C_i ($i = 1, 2$) is a 3-ball. Hence M is the 3-sphere and therefore M is irreducible, a contradiction. Hence we may assume that $\mathrm{genus}(S) > 0$. Let P be a ∂-reducing disk of M with $|P \cap S| = 1$. Changing subscripts, if necessary, we may assume that $P \cap C_1 = D$ is a disk and $P \cap C_2 = A$ is a spanning annulus.

Suppose that $\mathrm{genus}(S) = 1$. Since C_i ($i = 1, 2$) are irreducible, we see

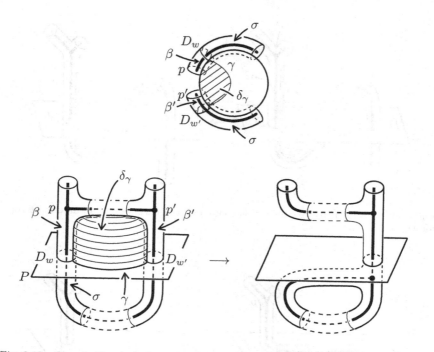

Fig. 3.13   Case 3. If $p \neq p'$, then β is also isotoped along δ_γ so that $|P \cap \Sigma|$ is reduced.

that ∂C_i contain no 2-sphere components. Since C_1 contains an essential disk D, we see that C_1 is homeomorphic to $\mathbb{B}^2 \times \mathbb{S}^1$. Since C_2 contains a spanning annulus A, we see that C_2 is homeomorphic to $\mathbb{T}^2 \times [0, 1]$. It follows that M is homeomorphic to $\mathbb{B}^2 \times \mathbb{S}^1$ and hence M is irreducible, a contradiction.

Suppose that genus$(S) > 1$. Let C_1' (C_2' resp.) be the manifold obtained from C_1 (C_2 resp.) by cutting along D (A resp.), and let A^+ and A^- be copies of A in $\partial C_2'$. Then we see that C_1' consists of either a single compression body or a union of two compression bodies (cf. (6) of Remark 2.1.4). Let C_2'' be the manifold obtained from C_2' by attaching 2-handles along A^+ and A^-. It follows from Remark 2.1.7 that C_2'' consists of either a compression body or a union of two compression bodies.

Suppose that C_1' consists of a single compression body. This implies that C_2'' consists of a single compression body (cf. Fig. 3.16). We can naturally obtain a homeomorphism $\partial_+ C_1' \to \partial_+ C_2''$ from the homeomorphism

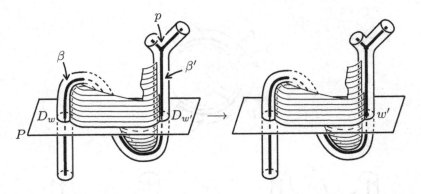

Fig. 3.14   Case 3. If $p = p'$, then β' is contained in β because γ is an outermost edge for D_w of Γ.

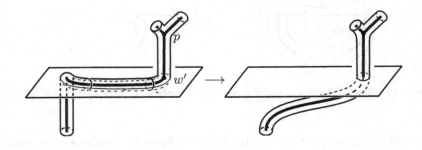

Fig. 3.15   Case 3. If $p = p'$, then $\mathrm{cl}(\sigma \setminus \beta')$ is isotoped so that $|P \cap \Sigma|$ is reduced.

$\partial_+ C_1 \to \partial_+ C_2$. Set $\partial_+ C_1' = \partial_+ C_2'' = S'$. Then $(C_1', C_2''; S')$ is a Heegaard splitting of the 3-manifolds M' obtained by cutting M along P. Note that $\mathrm{genus}(S') = \mathrm{genus}(S) - 1$. Moreover, it follows from an innermost disk argument that M' is also reducible.

Claim.

(1) If C_1' is reducible, then C_1 is reducible.
(2) If C_2'' is reducible, then one of the following holds.

    (a) C_2 is reducible.
    (b) The component of $\partial_- C_2$ intersecting A is a torus, say T.

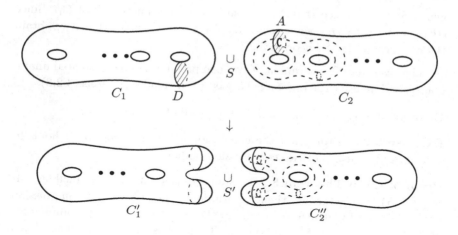

Fig. 3.16 If C_1' consists of a single compression body, so does C_2''.

Proof. Exercise 3.2.7. □

Recall that we assume that C_i ($i = 1, 2$) are irreducible. Hence it follows from (1) of the claim that C_1' is irreducible. It also follows from (2) of the claim that either (i) C_2'' is irreducible or (ii) C_2'' is reducible and the condition (b) of (2) in Claim 1 holds.

Suppose that the condition (i) holds. Then by induction on the genus of a Heegaard surface, $(C_1', C_2''; S')$ is reducible, *i.e.*, there are meridian disks D_1 and D_2 of C_1' and C_2'' respectively with $\partial D_1 = \partial D_2$. Note that this implies that C_i ($i = 1, 2$) are non-trivial. Let α^+ and α^- be the co-cores of the 2-handles attached to C_2''. Then we see that $\alpha^+ \cup \alpha^-$ is vertical in C_2''. It follows from Lemma 2.1.9 that we may assume that $D_2 \cap (\alpha^+ \cup \alpha^-) = \emptyset$, *i.e.*, D_2 is disjoint from the 2-handles. Hence the pair of disks D_1 and D_2 survives when we restore C_1 and C_2 from C_1' and C_2'' respectively. This implies that $(C_1, C_2; S)$ is reducible and hence we obtain the conclusion (1) of Theorem 3.2.1.

Suppose that the condition (ii) holds. Then it follows from (2) of Remark 2.1.7 that there is a separating disk E_2 in C_2 such that E_2 is disjoint from A and that E_2 cuts off $T^2 \times [0, 1]$ from C_2 with $T^2 \times \{0\} = T$. Let ℓ be a simple closed curve in $S \cap (T^2 \times \{1\})$ which intersects $A \cap S (= \partial A \cap S = \partial D)$ in a single point. Let E_1 be a disk properly embedded in C_1 which is obtained from two parallel copies of D by adding a band along ℓ. Since genus$(S) > 1$, we see that E_1 is a separating meridian disk of C_1. Since ∂E_1 is isotopic to ∂E_2, we see that $(C_1, C_2; S)$ is reducible. Hence we obtain the conclusion (1) of Theorem 3.2.1.

The case that C'_1 is a union of two compression bodies is treated analogously, and we leave the proof for this case to the reader (Exercise 3.2.8). $\square$

Exercise 3.2.7. Show the claim in the proof of Theorem 3.2.1.

Exercise 3.2.8. Prove that the conclusion (1) of Theorem 3.2.1 holds in case that C'_1 consists of two compression bodies.

Proof of (2) in Theorem 3.2.1. Suppose that M is ∂-reducible. If $(C_1, C_2; S)$ is ∂-reducible, then we are done. Let $\widehat{C}_i$ be the compression body obtained by attaching 3-balls to the 2-sphere boundary components of C_i $(i = 1, 2)$. Set $\widehat{M} = \widehat{C}_1 \cup \widehat{C}_2$. Then $\widehat{M}$ is also ∂-reducible. Then it follows from (1) of Remark 2.1.4 and Proposition 3.2.2 that $(\widehat{C}_1, \widehat{C}_2; S)$ is reducible or ∂-reducible. If $(\widehat{C}_1, \widehat{C}_2; S)$ is ∂-reducible, then we see that $(C_1, C_2; S)$ is also ∂-reducible. Hence we may assume that $(\widehat{C}_1, \widehat{C}_2; S)$ is reducible. By induction on the genus of a Heegaard surface, we prove that $(C_1, C_2; S)$ is ∂-reducible. Let P' be a reducing 2-sphere of $\widehat{M}$ with $|P' \cap S| = 1$. For each $i = 1$ and 2, set $D_i = P' \cap \widehat{C}_i$, and let $\widehat{C}'_i$ be the manifold obtained by cutting $\widehat{C}_i$ along D_i, and let D_i^+ and D_i^- be copies of D_i in $\partial \widehat{C}'_i$. Then each of $\widehat{C}'_i$ $(i = 1, 2)$ is either (1) a compression body if D_i is non-separating in $\widehat{C}_i$ or (2) a union of two compression bodies if D_i is separating in $\widehat{C}_i$. Note that we can naturally obtain a homeomorphism $\partial_+ \widehat{C}'_1 \to \partial_+ \widehat{C}'_2$ from the homeomorphism $\partial_+ \widehat{C}_1 \to \partial_+ \widehat{C}_2$. Set $\widehat{M}' = \widehat{C}'_1 \cup \widehat{C}'_2$ and $\partial_+ \widehat{C}'_1 = \partial_+ \widehat{C}'_2 = S'$. Then $(\widehat{C}'_1, \widehat{C}'_2; S')$ is either (1) a Heegaard splitting or (2) a union of two Heegaard splittings (cf. Fig. 3.17).

It follows from an innermost disk argument that there is a ∂-reducing disk of $\widehat{M}$ disjoint from P'. This implies that a component of $\widehat{M}'$ is ∂-reducible and hence one of the Heegaard splittings of $(\widehat{C}'_1, \widehat{C}'_2; S')$ is ∂-reducible. By induction on the genus of a Heegaard surface, we see that $(\widehat{C}_1, \widehat{C}_2; S)$ is ∂-reducible. Therefore $(C_1, C_2; S)$ is also ∂-reducible and hence we have (2) of Theorem 3.2.1. $\square$

Note that the following holds as a corollary of (2) of Theorem 3.2.1 together with Lemma 3.1.12.

Corollary 3.2.9. *Let $(C_1, C_2; S)$ be a non-trivial Heegaard splitting of $(M; \partial_1 M, \partial_2 M)$. If M is ∂-reducible, then $(C_1, C_2; S)$ is weakly reducible.*

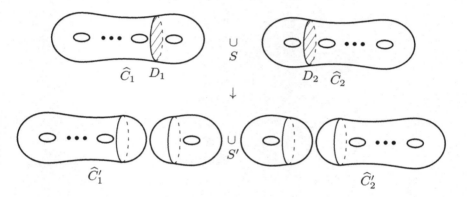

Fig. 3.17   The case that $(\widehat{C}_1', \widehat{C}_2'; S')$ is a union of two Heegaard splittings

Chapter 4

Two theorems on Heegaard splittings

Most of this chapter is devoted to a theorem of Waldhausen. For Waldhausen's original work, see [Waldhausen (1968)]. We also take a look at the theorem of Reidemeister and Singer.

4.1 Waldhausen's Theorem

We devote this section to a simplified proof of the following theorem originally due to Waldhausen [Waldhausen (1968)]. To prove the theorem, we exploit Gabai's idea of "thin position" [Gabai (1987)], Johannson's technique [Johannson (1991)] and Otal's idea [Otal (1991)] of viewing the Heegaard splittings as a graph in three dimensional space.

Theorem 4.1.1 (Waldhausen's Theorem). *Any Heegaard splitting of* $\mathbb{S}^3$ *is standard, i.e., obtained from the trivial Heegaard splitting by stabilization.*

To underline the extent of this theorem, we reiterate that every tunnel system of every knot in $\mathbb{S}^3$ gives rise to a Heegaard splitting of $\mathbb{S}^3$.

Thin position of graphs in the 3-sphere. Let $\Gamma \subset \mathbb{S}^3$ be a finite graph in which all vertices are of valence three. Let $h : \mathbb{S}^3 \to [-1, 1]$ be a height function such that $h^{-1}(t) = P(t) \cong \mathbb{S}^2$ for $t \in (-1, 1)$, $h^{-1}(-1) = $ (the south pole of $\mathbb{S}^3$), and $h^{-1}(1) = $ (the north pole of $\mathbb{S}^3$). Let $\mathcal{V}$ denote a set of the vertices of Γ.

Definition 4.1.2. The graph Γ is in *Morse position* with respect to h if the following conditions are satisfied.

(1) $h|_{\Gamma \setminus \mathcal{V}}$ has finitely many non-degenerate critical points.

(2) The height of critical points of $h|_{\Gamma \setminus \mathcal{V}}$ and of the elements of $\mathcal{V}$ are mutually different.

A set of the *critical heights* for Γ is the set of height at which there is either a critical point of $h|_{\Gamma \setminus \mathcal{V}}$ or an element of $\mathcal{V}$. We can deform Γ by an isotopy so that a regular neighborhood of each vertex v of Γ is either of *Type-y* (*i.e.*, two edges incident to v is above v and the remaining edge is below v) or of *Type-λ* (*i.e.*, two edges incident to v is below v and the remaining edge is above v). Such a graph is said to be in *normal form*. We call a vertex v a *y-vertex* (a *λ-vertex* resp.) if $\eta(v; \Gamma)$ is of Type-y (Type-λ resp.).

Suppose that Γ is in Morse position and in normal form. Note that $\eta(\Gamma; \mathbb{S}^3)$ can be regarded as the union of 0-handles corresponding to the regular neighborhood of the vertices and 1-handles corresponding to the regular neighborhood of the edges. A simple closed curve α in $\partial \eta(\Gamma; \mathbb{S}^3)$ is in *normal form* if the following conditions are satisfied.

(a) For each 1-handle ($\cong \mathbb{B}^2 \times [0,1]$), each component of $\alpha \cap (\partial \mathbb{B}^2 \times [0,1])$ is an essential arc in the annulus $\partial \mathbb{B}^2 \times [0,1]$.

(b) For each 0-handle ($\cong \mathbb{B}^3$), each component of $\alpha \cap \partial \mathbb{B}^3$ is an arc which is essential in the 2-sphere with three holes $\mathrm{cl}(\partial \mathbb{B}^3 \setminus$ (the 1-handles incident to $\mathbb{B}^3$)).

Let D be a disk properly embedded in $\mathrm{cl}(\mathbb{S}^3 \setminus \eta(\Gamma; \mathbb{S}^3))$. We say that D is in *normal form* if the following conditions are satisfied.

(1) ∂D is in normal form.

(2) Each critical point of $h|_{\mathrm{int}(D)}$ is non-degenerate.

(3) No critical points of $h|_{\mathrm{int}(D)}$ occur at critical heights of Γ.

(4) No two critical points of $h|_{\mathrm{int}(D)}$ occur at the same height.

(5) $h|_{\partial D}$ is a Morse function on ∂D satisfying the following (cf. Fig. 4.1).

   (a) Each minimum of $h|_{\partial D}$ occurs at a y-vertex in "half-center" singularity, at a minimum of Γ in "half-center" singularity, or at a λ-vertex in "half-saddle" singularity.

   (b) Each maximum of $h|_{\partial D}$ occurs either at a λ-vertex in "half-center" singularity, at a maximum of Γ in "half-center" singularity, or at a y-vertex in "half-saddle" singularity.

It is known by Morse theory that D can be put in normal form. See [Milnor (1963)]. Recall that $h : \mathbb{S}^3 \to [-1,1]$ is a height function such that $h^{-1}(t) = P(t) \cong \mathbb{S}^2$ for $t \in (-1,1)$, $h^{-1}(-1) =$ (the south pole of $\mathbb{S}^3$), and

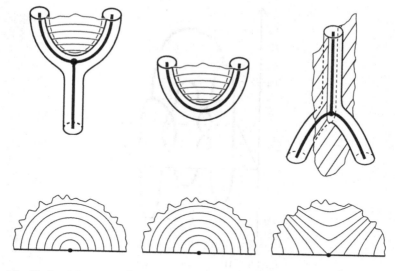

Fig. 4.1 Each minimum of $h|_{\partial D}$ occurs at a y-vertex in "half-center" singularity (left), at a minimum of Γ in "half-center" singularity (middle), or at a λ-vertex in "half-saddle" singularity (right).

$h^{-1}(1) =$ (the north pole of $\mathbb{S}^3$). We isotope Γ to be in Morse position and in normal form. For $t \in (-1, 1)$, set $w_\Gamma(t) = |P(t) \cap \Gamma|$. Note that $w_\Gamma(t)$ is constant on each component of $(-1, 1) \setminus$ (the critical heights of Γ). Set $W_\Gamma = \max\{w_\Gamma(t) | t \in (-1, 1)\}$ (cf. Fig. 4.2). Let n_Γ be the number of the components of $(-1, 1) \setminus$ (the critical heights of Γ) on which the value W_Γ is attained.

Definition 4.1.3. A graph $\Gamma \subset \mathbb{S}^3$ is said to be in *thin position* if (W_Γ, n_Γ) is minimal with respect to lexicographic order among all graphs which are obtained from Γ by ambient isotopies and edge slides and are in Morse position and in normal form.

We first prove Theorem 4.1.1 by using the proposition below. Let $(C_1, C_2; S)$ be a genus $g > 0$ Heegaard splitting of $\mathbb{S}^3$. Let Σ be a trivalent spine of C_1. Note that $\eta(\partial_- C_1 \cup \Sigma; M)$ is obtained from regular neighborhoods of $\partial_- C_1$ and the vertices of Σ by attaching 1-handles corresponding to the edges of Σ. Set $\Sigma_\eta = \eta(\Sigma; M)$. As in Section 3.2, the notation h_v^0, called a *vertex* of Σ_η, means a regular neighborhood of a vertex v of Σ. Also, the notation h_σ^1, called an *edge* of Σ_η, means a 1-handle correspond-

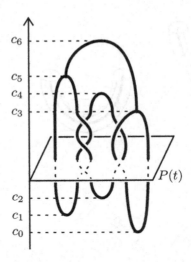

Fig. 4.2 $W_\Gamma = 6$ in this example.

ing to an edge σ of Σ. Let Δ_1 (Δ_2 resp.) be a complete meridian system of C_1 (C_2 resp.).

Proposition 4.1.4. *There is an edge of Σ_η which is disjoint from Δ_2, or Σ is modified by edge slides so that the resulting graph contains an unknotted cycle, i.e., the modified graph contains a graph α so that α bounds a disk in $\mathbb{S}^3$.*

Proof of Theorem 4.1.1 via Proposition 4.1.4. We prove Theorem 4.1.1 by induction on the genus of a Heegaard surface. If genus(S) = 0, then $(C_1, C_2; S)$ is standard (cf. Definition 3.1.10). So we may assume that genus(S) > 0 for a Heegaard splitting $(C_1, C_2; S)$.

Suppose first that Σ has an unknotted cycle α. Then $\eta(\alpha; C_1)$ is a standard solid torus in $\mathbb{S}^3$, that is, the exterior of $\eta(\alpha; C_1)$ is a solid torus. Since $C_1^- = \mathrm{cl}(C_1 \setminus \eta(\alpha; C_1))$ is a compression body, we see that $(C_1^-, C_2; S)$ is a Heegaard splitting of $\mathrm{cl}(\mathbb{S}^3 \setminus \eta(\alpha; C_1))$ which is a solid torus. Since a solid torus is ∂-reducible, it follows from Theorem 3.2.1 that $(C_1^-, C_2; S)$ is ∂-reducible, that is, there is a ∂-reducing disk D_α for $(C_1^-, C_2; S)$ with $|D_\alpha \cap S| = 1$. Since $\eta(\alpha; C_1)$ is a standard solid torus in $\mathbb{S}^3$, we see that D_α intersects a meridian disk D'_α of $\eta(\alpha; C_1)$ transversely in a single point. Set $D_2 = D_\alpha \cap C_2$. Extending D'_α naturally, we obtain a meridian disk D_1 of C_1 such that ∂D_1 intersects ∂D_2 transversely in a single point, i.e., D_1

and D_2 give stabilization of $(C_1, C_2; S)$ (cf. Fig. 4.3). Hence we obtain a Heegaard splitting $(C_1', C_2'; S')$ with genus$(S') <$ genus(S). By induction on the genus of a Heegaard surface, we can see that $(C_1, C_2; S)$ is standard.

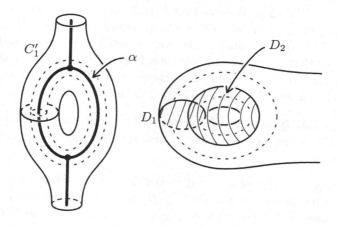

Fig. 4.3   A pair of (D_1, D_2) gives a stabilization of $(C_1, C_2; S)$.

Suppose next that there is an edge σ of Σ with $h_\sigma^1 \cap \Delta_2 = \emptyset$. Let D_σ be a meridian disk of C_1 which is co-core of the 1-handle h_σ^1. Note that $D_\sigma \cap \Delta_2 = \emptyset$. Cutting C_2 along Δ_2, we obtain a union of 3-balls and hence we see that ∂D_σ bounds a disk, say D_σ', properly embedded in one of the 3-balls. Note that D_σ' corresponds to a meridian disk of C_2. Hence we see that $(C_1, C_2; S)$ is reducible. It follows from a generalized Schönflies theorem that every 2-sphere in $\mathbb{S}^3$ separates it into two 3-balls (cf. Section 2.F.5 of [Rolfsen (1990)]). Hence by cutting $\mathbb{S}^3$ along the reducing 2-sphere and capping off 3-balls, we obtain two Heegaard splittings of $\mathbb{S}^3$ such that the genus of each Heegaard surface is less than that of S. Then we see that $(C_1, C_2; S)$ is standard by induction on the genus of a Heegaard surface. $\square$

In the remainder of this section, we give a proof of Proposition 4.1.4. Let $h : \mathbb{S}^3 \to [-1, 1]$ be a height function such that $h^{-1}(t) = P(t) \cong \mathbb{S}^2$ for $t \in (-1, 1)$, $h^{-1}(-1) = $ (the south pole of $\mathbb{S}^3$), and $h^{-1}(1) = $ (the north pole of $\mathbb{S}^3$). We may assume that Σ is in thin position. We also assume that each component of Δ_2 is in normal form, Δ_1 intersects Δ_2 transversely and $|\Delta_1 \cap \Delta_2|$ is minimal.

For the proof of Proposition 4.1.4, it is enough to show the following; if there are no edges of Σ_η which are disjoint from Δ_2, then Σ is modified by edge slides so that the resulting graph contains an unknotted cycle. Hence we suppose that there are no edges of Σ_η which are disjoint from Δ_2. Set $\Lambda(t) = P(t) \cap (\Sigma_\eta \cup \Delta_2)$. We note that $P(t)$, Σ and Δ_2 intersect transversely at a regular height t. In the following, we mainly consider such a regular height t with $\Lambda(t) \neq \emptyset$ unless otherwise denoted. We also note that we may assume that $\Lambda(t)$ does not contain a loop component by an argument similar to the proof of Lemma 3.2.3. Hence $\Lambda(t)$ is regarded as a graph in $P(t)$ which consists of *fat-vertices* $P(t) \cap \Sigma_\eta$ and edges $P(t) \cap \Delta_2$.

Lemma 4.1.5. *If there is a fat-vertex of $\Lambda(t)$ with valence less than two, then Σ is modified by edge slides so that the resulting graph contains an unknotted cycle.*

Proof. Suppose that there is a fat-vertex D_v of $\Lambda(t)$ with valence less than two. Let σ be the edge of Σ with $h^1_\sigma \supset D_v$ and p one of the endpoints of σ. Since we assume that there are no edges of Σ_η which are disjoint from Δ_2, we see that any fat-vertex of $\Lambda(t)$ is of valence greater than zero. Hence D_v is of valence one. Then there is the disk component D of Δ_2 with $h^1_\sigma \cap D \neq \emptyset$. Since ∂D intersects the fat-vertex D_v in a single point and hence ∂D intersects h^1_σ in a single arc, we can perform an edge slide on σ along $\mathrm{cl}(\partial D \setminus h^1_\sigma)$ to obtain a new graph Σ' from Σ (cf. Fig. 4.4). Clearly, Σ' contains an unknotted cycle (bounding a disk corresponding to D_2). $\quad\square$

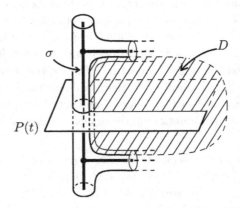

Fig. 4.4 Since ∂D intersects h^1_σ in a single arc, σ is isotoped along $\mathrm{cl}(\partial D \setminus h^1_\sigma)$ so that a new graph Σ' is obtained.

An edge of a graph $\Lambda(t)$ is said to be *simple* if the edge joins distinct two fat-vertices of $\Lambda(t)$. Recall that an edge of a graph $\Lambda(t)$ is called a *loop* if the edge is not simple.

Lemma 4.1.6. *Suppose that there are no fat-vertices of valence less than two. Then there exists a fat-vertex D_w of $\Lambda(t)$ such that any outermost edge for D_w of $\Lambda(t)$ is simple.*

Proof. If $\Lambda(t)$ does not contain a loop, then we are done. So we may assume that $\Lambda(t)$ contains a loop, say μ. Let D_v be a fat-vertex of $\Lambda(t)$ which is a bese of μ. It follows from an argument similar to the proof of Lemma 3.2.4 that μ cuts $\mathrm{cl}(P(t) \setminus D_v)$ into two disks, and each of the two disks contains a fat-vertex of $\Lambda(t)$. Let μ_0 be a loop of $\Lambda(t)$ which is innermost in $P(t)$. Let D_w be a fat-vertex contained in the interior of the innermost disk bounded by μ_0. Note that D_w is not isolated and that every edge contained in the interior of the innermost disk is simple. Hence any outermost edge for D_w of $\Lambda(t)$ is simple. $\qquad\square$

Let D_w be a fat-vertex of $\Lambda(t)$ with a simple edge $\gamma(\subset \Lambda(t))$. We may assume that γ is a simple outermost edge for D_w of $\Lambda(t)$ and γ is contained in a disk component D of Δ_2. It follows from Lemma 4.1.6 that we can always find such a fat-vertex D_w and an edge γ if each fat-vertex of $\Lambda(t)$ is of valence greater than one. Let δ_γ be an outermost disk for (D_w, γ). We say an outermost edge γ is *upper* (*lower* resp.) if $\eta(\gamma; \delta_\gamma)$ is above (below resp.) γ with respect to the height function h. Let t_0 be a regular height with $w_\Sigma(t_0) = W_\Sigma$.

Lemma 4.1.7. *Let D_w be a fat-vertex of $\Lambda(t_0)$ with a simple outermost edge for D_w of $\Lambda(t)$. Then we have one of the following.*

(1) All the simple outermost edges for D_w of $\Lambda(t)$ are either upper or lower.
(2) Σ is modified by edge slides so that the resulting graph contains an unknotted cycle.

Proof. Suppose that $\Lambda(t)$ contains simple outermost edges for D_w, say γ and γ', such that γ is upper and γ' is lower. For the proof of Lemma 4.1.7, it is enough to show that Σ is modified by edge slides so that there is an unknotted cycle. Let δ_γ and $\delta_{\gamma'}$ be the outermost disk for (D_w, γ) and $(D_{w'}, \gamma')$ respectively. Let σ be the edge of Σ with $h_\sigma^1 \supset D_w$. Let $\bar{\gamma}$ ($\bar{\gamma}'$ resp.) be a union of the components obtained by cutting σ by the two fat-vertices of $\Lambda(t_0)$ incident to γ (γ' resp.) such that a 1-handle corresponding to each component intersects $\partial\delta_\gamma \setminus \gamma$ ($\partial\delta_{\gamma'} \setminus \gamma'$ resp.). We note that $\bar\gamma$ ($\bar\gamma'$ resp.) satisfies one of the following conditions.

(1) $\bar\gamma$ ($\bar\gamma'$ resp.) consists of an arc such that $\bar\gamma$ ($\bar\gamma'$ resp.) and σ share a single endpoint.

(2) $\bar\gamma$ ($\bar\gamma'$ resp.) consists of an arc with $\bar\gamma \subset \mathrm{int}(\sigma)$ ($\bar\gamma' \subset \mathrm{int}(\sigma)$ resp.).

(3) $\bar\gamma$ ($\bar\gamma'$ resp.) consists of two subarcs of σ such that each component of $\bar\gamma$ ($\bar\gamma'$ resp.) and σ share a single endpoint.

In each of the conditions above, corresponding figures are illustrated in Fig. 4.5. (We remark that there is a case in the condition (3) which is similar to the latter half of Case 3 in the proof of Lemma 3.2.6. Recall that we only have to use broken edge slides in that case. Hence we omit details in such a case here.)

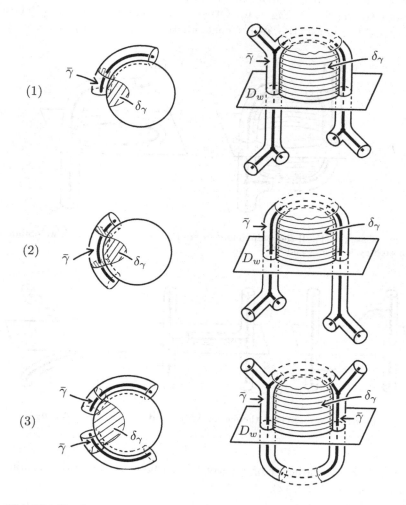

Fig. 4.5 Simple outermost edges $\bar{\gamma}$ and $\bar{\gamma}'$ satisfy one of the conditions above.

Case (1)-(1).   Both $\bar{\gamma}$ and $\bar{\gamma}'$ satisfy the condition (1).

If the endpoints of γ are the same as those of γ', then we can slide $\bar{\gamma}$ ($\bar{\gamma}'$ resp.) to γ (γ' resp.) along the disk δ_γ ($\delta_{\gamma'}$ resp.) and hence we obtain an unknotted cycle (cf. Fig. 4.6). Otherwise, we can perform a Whitehead move on Σ to reduce (W_Σ, n_Σ), a contradiction (cf. Fig. 4.6).

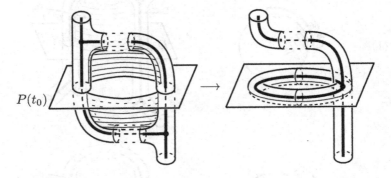

Fig. 4.6   Case (1)-(1). If the endpoints of γ are the same as those of γ', we obtain an unknotted cycle from $\bar{\gamma}$ and $\bar{\gamma}'$.

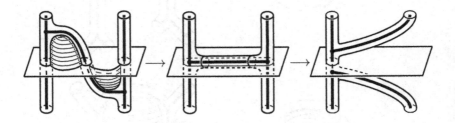

Fig. 4.7   Case (1)-(1). Otherwise, a Whitehead move reduces (W_Σ, n_Σ).

Case (1)-(2).   Either $\bar{\gamma}$ or $\bar{\gamma}'$, say $\bar{\gamma}$, satisfies the condition (1) and $\bar{\gamma}'$ satisfies the condition (2).

Then we can slide $\bar{\gamma}$ ($\bar{\gamma}'$ resp.) to γ (γ' resp.) along the disk δ_γ ($\delta_{\gamma'}$ resp.). Then Σ is further isotoped to reduce (W_Σ, n_Σ), a contradiction (cf. Fig. 4.8).

Case (1)-(3).   Either $\bar{\gamma}$ or $\bar{\gamma}'$, say $\bar{\gamma}$, satisfies the condition (1) and $\bar{\gamma}'$ satisfies the condition (3).

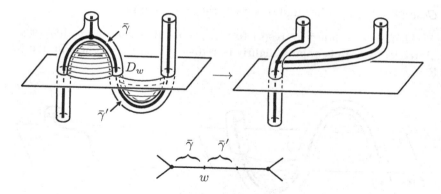

Fig. 4.8   Case (1)-(2). The edges $\bar{\gamma}$ and $\bar{\gamma}'$ are isotoped along δ_γ and $\delta_{\gamma'}$ respectively so that (W_Σ, n_Σ) is reduced.

Let $\bar{\gamma}_1'$ and $\bar{\gamma}_2'$ be the components of $\bar{\gamma}'$ with $h^1_{\bar{\gamma}_1'} \supset D_w$. Note that $\bar{\gamma} \supset \bar{\gamma}_2'$ and hence $\mathrm{int}(\bar{\gamma}) \supset \partial\bar{\gamma}_2'$. This implies that $\mathrm{int}(\bar{\gamma}) \cap P(t_0) \neq \emptyset$. Hence we can slide $\bar{\gamma}$ to γ along the disk δ_γ. We can further isotope Σ slightly to reduce (W_Σ, n_Σ), a contradiction (cf. Fig. 4.9).

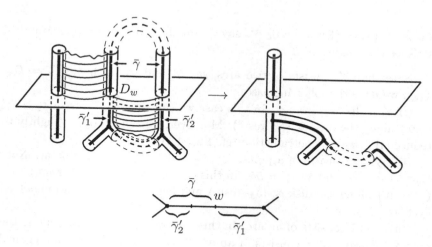

Fig. 4.9   Case (1)-(3). The edge $\bar{\gamma}$ is isotoped along δ_γ so that (W_Σ, n_Σ) is reduced.

Case (2)-(2).   Both $\bar{\gamma}$ and $\bar{\gamma}'$ satisfy the condition (2).

Then we can slide $\bar{\gamma}$ ($\bar{\gamma}'$ resp.) to γ (γ' resp.) along the disk δ_γ ($\delta_{\gamma'}$ resp.).
Moreover, we can isotope σ slightly to reduce (W_Σ, n_Σ), a contradiction (cf.
Fig. 4.10).

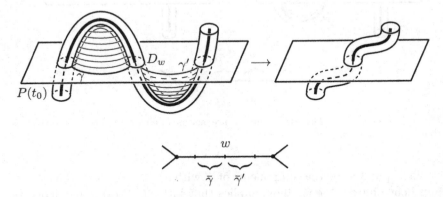

Fig. 4.10   Case (2)-(2). The edges $\bar{\gamma}$ and $\bar{\gamma}'$ are isotoped along δ_γ and $\delta_{\gamma'}$ respectively
so that (W_Σ, n_Σ) is reduced.

Case (2)-(3).   Either $\bar{\gamma}$ or $\bar{\gamma}'$, say $\bar{\gamma}$, satisfies the condition (2) and $\bar{\gamma}'$
satisfies the condition (3).

Note that $\bar{\gamma}'$ consists of two arcs, say $\bar{\gamma}'_1$ and $\bar{\gamma}'_2$, with $\bar{\gamma}'_1 \cap \bar{\gamma} = D_w$.
Then we have the following cases.

(i) $\bar{\gamma}'_2$ is disjoint from $\bar{\gamma}$. In this case we can slide $\bar{\gamma}$ ($\bar{\gamma}'$ resp.) to γ (γ'
resp.) along the disk δ_γ ($\delta_{\gamma'}$ resp.). Moreover, we can isotope σ slightly to
reduce (W_Σ, n_Σ), a contradiction (cf. Fig. 4.11).

(ii) $\bar{\gamma}'_2 \cap \bar{\gamma}$ consists of a point, *i.e.*, $\bar{\gamma}'_2$ and $\bar{\gamma}$ share one endpoint. Note
that $\bar{\gamma} \cup \bar{\gamma}' = \sigma$ and $\bar{\gamma} \cap \bar{\gamma}' = \partial\bar{\gamma}$. In this case we can slide $\bar{\gamma}$ ($\bar{\gamma}'$ resp.) to γ
(γ' resp.) along the disk δ_γ ($\delta_{\gamma'}$ resp.) and hence we obtain an unknotted
cycle (cf. Fig. 4.12).

(iii) $\bar{\gamma}'_2 \cap \bar{\gamma}$ consists of an arc. In this case we can slide $\bar{\gamma}$ to γ along the
disk δ_γ. Since $\bar{\gamma}'_2 \cap \bar{\gamma}$ consists of an arc, $\bar{\gamma}$ contains at least three critical
points. Hence we can further isotope σ slightly to reduce (W_Σ, n_Σ), a
contradiction (cf. Fig. 4.13).

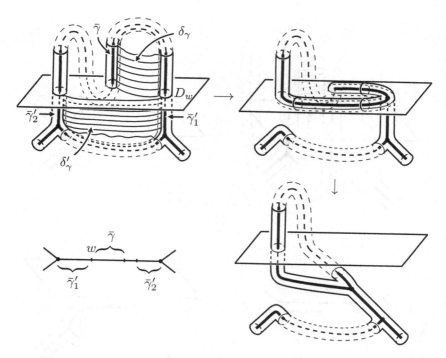

Fig. 4.11 Case (2)-(3). If $\bar{\gamma}_2'$ is disjoint from $\bar{\gamma}$, the edges $\bar{\gamma}$ and $\bar{\gamma}'$ are isotoped along δ_γ and $\delta_{\gamma'}$ respectively so that (W_Σ, n_Σ) is reduced.

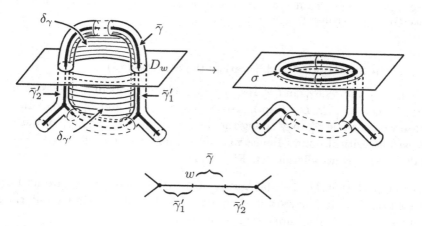

Fig. 4.12 Case (2)-(3). If $\bar{\gamma}_2' \cap \bar{\gamma}$ consists of a point, we obtain an unknotted cycle from $\bar{\gamma}_2'$ and $\bar{\gamma}$.

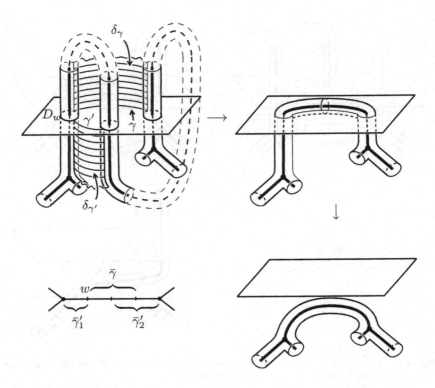

Fig. 4.13 Case (2)-(3). If $\bar{\gamma}_2' \cap \bar{\gamma}$ consists of an arc, the edge $\bar{\gamma}$ is isotoped along δ_γ so that (W_Σ, n_Σ) is reduced.

Case (3)-(3).   Both $\bar{\gamma}$ and $\bar{\gamma}'$ satisfy the condition (3).

Let $\bar{\gamma}_1$ and $\bar{\gamma}_2$ ($\bar{\gamma}_1'$ and $\bar{\gamma}_2'$ resp.) be the components of $\bar{\gamma}$ ($\bar{\gamma}'$ resp.) with $h_{\bar{\gamma}_1}^1 \supset D_w$ ($h_{\bar{\gamma}_1'}^1 \supset D_w$ resp.). Note that $\bar{\gamma}_1 \supset \bar{\gamma}_2'$ and $\bar{\gamma}_1' \supset \bar{\gamma}_2$. In this case we can slide $\bar{\gamma}_1'$ to γ' along the disk $\delta_{\gamma'}$. Since $\bar{\gamma}_1' \supset \bar{\gamma}_2$, $\bar{\gamma}_1'$ contains at least one critical point. Hence we can further isotope σ slightly to reduce (W_Σ, n_Σ), a contradiction (cf. Fig. 4.14).   □

Suppose that D_w is a fat-vertex of $\Lambda(t_0)$ such that there are no loops based on D_w. It follows from Lemma 4.1.7 that all the simple outermost edges for D_w of $\Lambda(t_0)$ are either upper or lower.

Lemma 4.1.8. *Suppose that all of the simple outermost edges for D_w of $\Lambda(t_0)$ are upper (lower resp.). Then one of the following holds.*

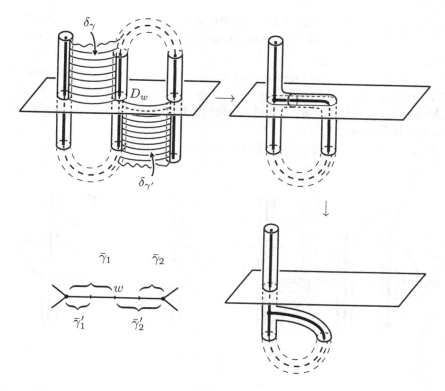

Fig. 4.14 Case (3)-(3). The edge $\bar{\gamma}'_1$ is isotoped along $\delta_{\gamma'}$ so that (W_Σ, n_Σ) is reduced.

(1) For each fat-vertex $D_{w'}$ of $\Lambda(t_0)$, every simple outermost edges for $D_{w'}$ of $\Lambda(t_0)$ is upper (lower resp.).

(2) Σ is modified by edge slides so that the modified graph contains an unknotted cycle.

Proof. Since the arguments are symmetric, we may suppose that all the simple outermost edges for D_w of $\Lambda(t_0)$ are upper. Let γ be a simple outermost edge for D_w of $\Lambda(t_0)$. Note that γ is upper. Suppose that there is a fat-vertex $D_{w'}$ such that $\Lambda(t_0)$ contains a lower simple outermost edge γ' for D_w. Let δ_γ ($\delta_{\gamma'}$ resp.) be the outermost disk for (D_w, γ) ($(D_{w'}, \gamma')$ resp.). Let σ (σ' resp.) be the edge of Σ with $h^1_\sigma \supset D_w$ ($h^1_{\sigma'} \supset D_{w'}$ resp.). Let $\bar{\gamma}$ ($\bar{\gamma}'$ resp.) be a union of the components obtained by cutting σ by the two fat-vertices of $\Lambda(t_0)$ incident to γ (γ' resp.) such that a 1-handle corresponding to each component intersects $\partial\delta_\gamma \setminus \gamma$ ($\partial\delta_{\gamma'} \setminus \gamma'$ resp.). Then

$\bar{\gamma}$ ($\bar{\gamma}'$ resp.) satisfies one of the conditions (1), (2) and (3) in the proof of Lemma 4.1.7. The proof of Lemma 4.1.8 is divided into the following cases.

Case A. $\quad \bar{\gamma} \cap \bar{\gamma}' = \emptyset$.

Then we have the following six cases. In each case, we can slide (a component of) $\bar{\gamma}$ ($\bar{\gamma}'$ resp.) to γ (γ' resp.) along the disk δ_γ ($\delta_{\gamma'}$ resp.). Moreover, we can isotope σ and σ' slightly to reduce (W_Σ, n_Σ) is reduced, a contradiction.

Case A-(1)-(1). Both $\bar{\gamma}$ and $\bar{\gamma}'$ satisfy the condition (1).

See Fig. 4.15.

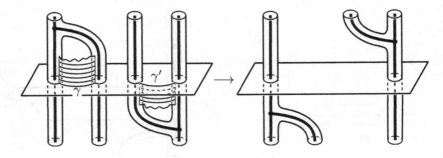

Fig. 4.15   Case A-(1)-(1).

Case A-(1)-(2). Either $\bar{\gamma}$ or $\bar{\gamma}'$, say $\bar{\gamma}$, satisfies the condition (1) and $\bar{\gamma}'$ satisfies the condition (2).

See Fig. 4.16.

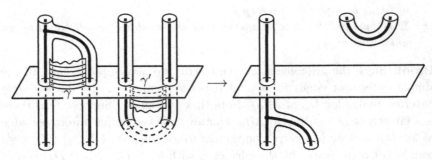

Fig. 4.16   Case A-(1)-(2).

Case A-(1)-(3).   Either $\bar{\gamma}$ or $\bar{\gamma}'$, say $\bar{\gamma}$, satisfies the condition (1) and $\bar{\gamma}'$ satisfies the condition (3).

See Fig. 4.17.

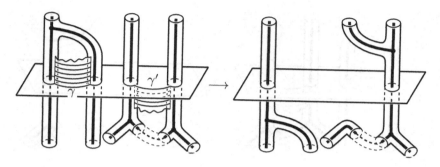

Fig. 4.17   Case A-(1)-(3).

Case A-(2)-(2).   Both $\bar{\gamma}$ and $\bar{\gamma}'$ satisfy the condition (2).

See Fig. 4.18.

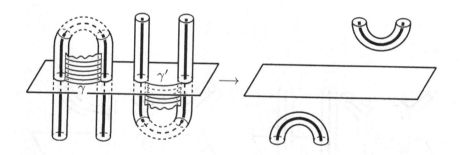

Fig. 4.18   Case A-(2)-(2).

Case A-(2)-(3).   Either $\bar{\gamma}$ or $\bar{\gamma}'$, say $\bar{\gamma}$, satisfies the condition (2) and $\bar{\gamma}'$ satisfies the condition (3).

See Fig. 4.19.

Case A-(3)-(3).   Both $\bar{\gamma}$ and $\bar{\gamma}'$ satisfy the condition (3).

See Fig. 4.20.

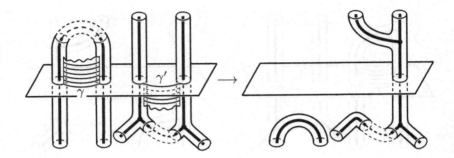

Fig. 4.19   Case A-(2)-(3).

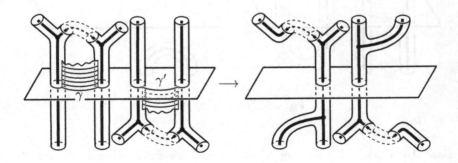

Fig. 4.20   Case A-(3)-(3).

Case B. $\bar{\gamma} \cap \bar{\gamma}' \neq \emptyset$.

Case B-(1)-(1).   Both $\bar{\gamma}$ and $\bar{\gamma}'$ satisfy the condition (1).

We first suppose that $\mathrm{int}(\bar{\gamma}) \cap \mathrm{int}(\bar{\gamma}') = \emptyset$. Then we can slide $\bar{\gamma}$ ($\bar{\gamma}'$ resp.) to γ (γ' resp.) along the disk δ_γ ($\delta_{\gamma'}$ resp.). If $\partial\gamma = \partial\gamma'(= \{w, w'\})$, then $\bar{\gamma} \cup \bar{\gamma}'$ composes an unknotted cycle and hence Lemma 4.1.8 holds (cf. Fig. 4.21). Otherwise, we can perform a Whitehead move on Σ and hence we can reduce (W_Σ, n_Σ), a contradiction (cf. Fig. 4.22).

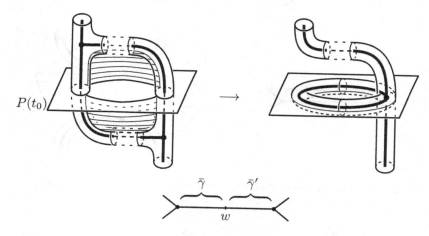

Fig. 4.21   Case B-(1)-(1). If $\mathrm{int}(\bar{\gamma}) \cap \mathrm{int}(\bar{\gamma}') = \emptyset$, we obtain an unknotted cycle from $\bar{\gamma}$ and $\bar{\gamma}'$.

We next suppose that $\mathrm{int}(\bar{\gamma}) \cap \mathrm{int}(\bar{\gamma}') \neq \emptyset$. Then there are two possibilities: (1) $\bar{\gamma} \subset \bar{\gamma}'$ or $\bar{\gamma}' \subset \bar{\gamma}$, say the latter holds and (2) $\bar{\gamma} \not\subset \bar{\gamma}'$ and $\bar{\gamma}' \not\subset \bar{\gamma}$. In each case, we can slide $\bar{\gamma}$ to γ along the disk δ_γ. Moreover, we can isotope σ slightly to reduce (W_Σ, n_Σ), a contradiction (cf. Fig. 4.23 and Fig. 4.24).

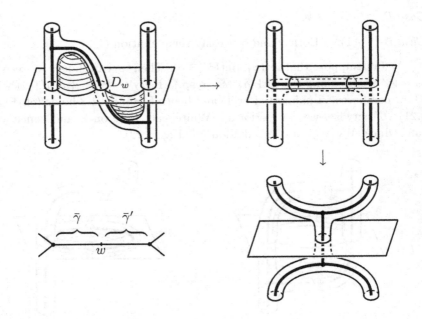

Fig. 4.22   Case B-(1)-(1). Otherwise, a Whitehead move reduces (W_Σ, n_Σ).

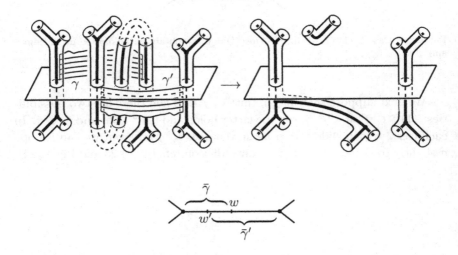

Fig. 4.23   Case B-(1)-(1). If $\operatorname{int}(\bar\gamma) \cap \operatorname{int}(\bar\gamma') \neq \emptyset$ and $\bar\gamma' \subset \bar\gamma$, the edge $\bar\gamma$ is isotoped along δ_γ so that (W_Σ, n_Σ) is reduced.

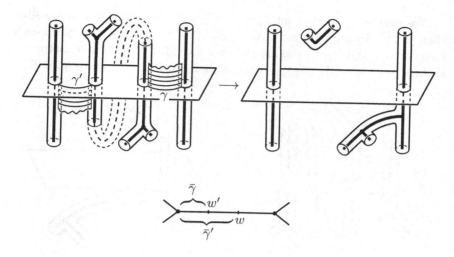

Fig. 4.24 Case B-(1)-(1). If $\text{int}(\bar{\gamma}) \cap \text{int}(\bar{\gamma}') \neq \emptyset$, $\bar{\gamma} \not\subset \bar{\gamma}'$ and $\bar{\gamma}' \not\subset \bar{\gamma}$, the edge $\bar{\gamma}$ is isotoped along δ_{γ} so that (W_{Σ}, n_{Σ}) is reduced.

Case B-(1)-(2).   Either $\bar{\gamma}$ or $\bar{\gamma}'$, say $\bar{\gamma}$, satisfies the condition (1) and $\bar{\gamma}'$ satisfies the condition (2).

We first suppose that $\text{int}(\bar{\gamma}) \cap \text{int}(\bar{\gamma}') = \emptyset$. Then we can slide $\bar{\gamma}$ ($\bar{\gamma}'$ resp.) to γ (γ' resp.) along the disk δ_{γ} ($\delta_{\gamma'}$ resp.). Moreover, we can isotope σ slightly to reduce (W_{Σ}, n_{Σ}), a contradiction (cf. Fig. 4.25).

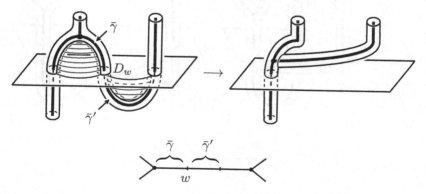

Fig. 4.25 Case B-(1)-(2). If $\text{int}(\bar{\gamma}) \cap \text{int}(\bar{\gamma}') = \emptyset$, the edges $\bar{\gamma}$ and $\bar{\gamma}'$ are isotoped along δ_{γ} and $\delta_{\gamma'}$ respectively so that (W_{Σ}, n_{Σ}) is reduced.

We next suppose that $\text{int}(\bar{\gamma}) \cap \text{int}(\bar{\gamma}') \neq \emptyset$. Note that it is impossible that $\bar{\gamma} \subset \bar{\gamma}'$. Hence there are two possibilities: $\bar{\gamma}' \subset \bar{\gamma}$ and $\bar{\gamma}' \not\subset \bar{\gamma}$. In each case, we can slide $\bar{\gamma}$ to γ along the disk δ_γ. Moreover, we can isotope σ slightly to reduce (W_Σ, n_Σ), a contradiction (cf. Fig. 4.26 and Fig. 4.27).

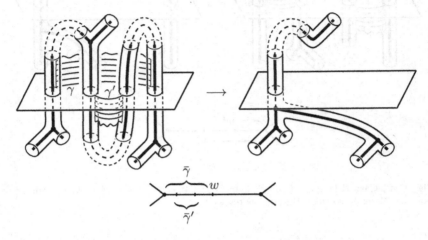

Fig. 4.26 Case B-(1)-(2). If $\text{int}(\bar{\gamma}) \cap \text{int}(\bar{\gamma}') \neq \emptyset$ and $\bar{\gamma}' \subset \bar{\gamma}$, the edge $\bar{\gamma}$ is isotoped along δ_γ so that (W_Σ, n_Σ) is reduced.

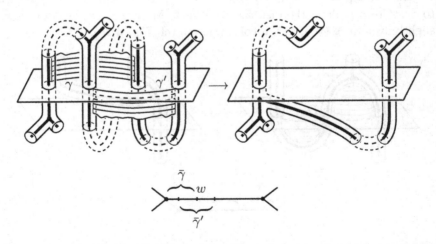

Fig. 4.27 Case B-(1)-(2). If $\text{int}(\bar{\gamma}) \cap \text{int}(\bar{\gamma}') \neq \emptyset$ and $\bar{\gamma}' \not\subset \bar{\gamma}$, the edge $\bar{\gamma}$ is isotoped along δ_γ so that (W_Σ, n_Σ) is reduced.

Case B-(1)-(3). Either $\bar\gamma$ or $\bar\gamma'$, say $\bar\gamma$, satisfies the condition (1) and $\bar\gamma'$ satisfies the condition (3).

Let $\bar\gamma_1'$ and $\bar\gamma_2'$ be the components of $\bar\gamma'$ with $h_{\bar\gamma_1'}^1 \supset D_{w'}$.

We first suppose that $\bar\gamma \subset \bar\gamma_1'$. Then we can slide $\bar\gamma_1'$ into γ' along the disk $\delta_{\gamma'}$. Moreover, we can isotope σ slightly to reduce (W_Σ, n_Σ), a contradiction (cf. Fig. 4.28).

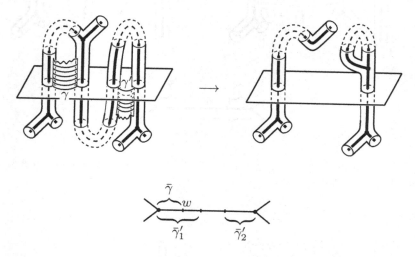

Fig. 4.28 Case B-(1)-(3). If $\bar\gamma \subset \bar\gamma_1'$, the edge $\bar\gamma_1'$ is isotoped along $\delta_{\gamma'}$ so that (W_Σ, n_Σ) is reduced.

We next suppose that $\bar\gamma_1' \subset \bar\gamma$. Then there are two possibilities: $\bar\gamma_2' \cap \bar\gamma = \emptyset$ and $\bar\gamma_2' \cap \bar\gamma \ne \emptyset$. In each case, we can slide $\bar\gamma$ to γ along the disk δ_γ. Moreover, we can isotope σ slightly to reduce (W_Σ, n_Σ), a contradiction (cf. Fig. 4.29 and Fig. 4.30).

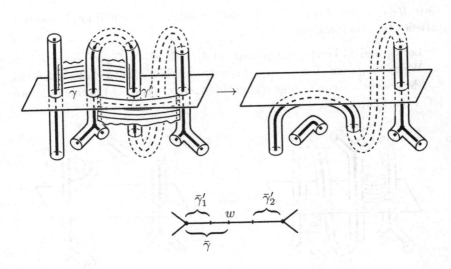

Fig. 4.29   Case B-(1)-(3). If $\bar{\gamma}'_1 \subset \bar{\gamma}$ and $\bar{\gamma}'_2 \cap \bar{\gamma} \neq \emptyset$, the edge $\bar{\gamma}$ is isotoped along δ_γ so that (W_Σ, n_Σ) is reduced.

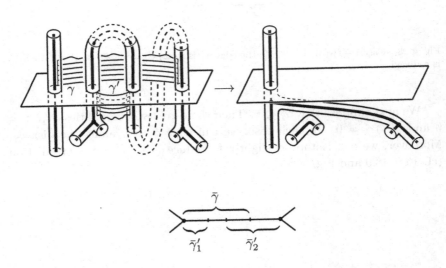

Fig. 4.30   Case B-(1)-(3). If $\bar{\gamma}'_1 \subset \bar{\gamma}$ and $\bar{\gamma}'_2 \cap \bar{\gamma} = \emptyset$, the edge $\bar{\gamma}$ is isotoped along δ_γ so that (W_Σ, n_Σ) is reduced.

Case B-(2)-(2).   Both $\bar{\gamma}$ and $\bar{\gamma}'$ satisfy the condition (2).

We first suppose that $\text{int}(\bar{\gamma}) \cap \text{int}(\bar{\gamma}') = \emptyset$. Then we can slide $\bar{\gamma}$ ($\bar{\gamma}'$ resp.) to γ (γ' resp.) along the disk δ_γ ($\delta_{\gamma'}$ resp.). Moreover, we can isotope σ slightly to reduce (W_Σ, n_Σ), a contradiction (cf. Fig. 4.31).

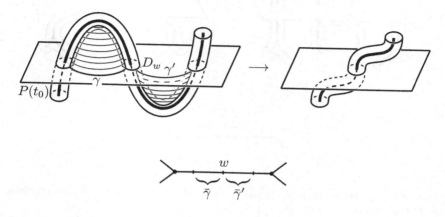

Fig. 4.31   Case B-(2)-(2). If $\text{int}(\bar{\gamma}) \cap \text{int}(\bar{\gamma}') = \emptyset$, the edges $\bar{\gamma}$ and $\bar{\gamma}'$ are isotoped along δ_γ and $\delta_{\gamma'}$ respectively so that (W_Σ, n_Σ) is reduced.

We next suppose that $\text{int}(\bar{\gamma}) \cap \text{int}(\bar{\gamma}') \neq \emptyset$. Then there are two possibilities: (1) $\bar{\gamma} \subset \bar{\gamma}'$ or $\bar{\gamma}' \subset \bar{\gamma}$, say the latter holds and (2) $\bar{\gamma} \not\subset \bar{\gamma}'$ and $\bar{\gamma}' \not\subset \bar{\gamma}$. In each case, we can slide $\bar{\gamma}$ to γ along the disk δ_γ. Moreover, we can isotope σ slightly to reduce (W_Σ, n_Σ), a contradiction (cf. Fig. 4.32 and Fig. 4.33).

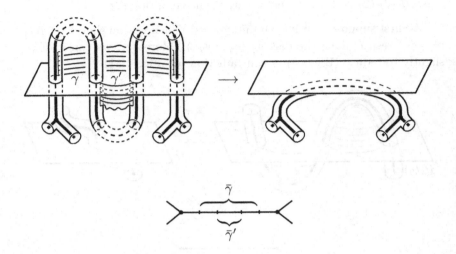

Fig. 4.32   Case B-(2)-(2). If $\operatorname{int}(\bar\gamma) \cap \operatorname{int}(\bar\gamma') \neq \emptyset$ and $\bar\gamma' \subset \bar\gamma$, the edge $\bar\gamma$ is isotoped along δ_γ so that (W_Σ, n_Σ) is reduced.

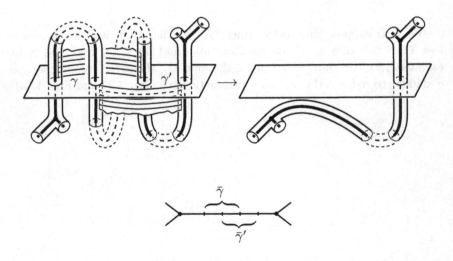

Fig. 4.33   Case B-(2)-(2). If $\operatorname{int}(\bar\gamma) \cap \operatorname{int}(\bar\gamma') \neq \emptyset$, $\bar\gamma \not\subset \bar\gamma'$ and $\bar\gamma' \not\subset \bar\gamma$, the edge $\bar\gamma$ is isotoped along δ_γ so that (W_Σ, n_Σ) is reduced.

Case B-(2)-(3).   Either $\bar{\gamma}$ or $\bar{\gamma}'$, say $\bar{\gamma}$, satisfies the condition (2) and $\bar{\gamma}'$ satisfies the condition (3).

Let $\bar{\gamma}_1'$ and $\bar{\gamma}_2'$ be the components of $\bar{\gamma}'$ with $\partial\bar{\gamma}_1' \supset D_{w'}$.

We first suppose that $\operatorname{int}(\bar{\gamma}) \cap \operatorname{int}(\bar{\gamma}') = \emptyset$. Since $\bar{\gamma} \cap \bar{\gamma}' \neq \emptyset$, we may suppose that $\bar{\gamma}_1' \cap \bar{\gamma}(= \partial\bar{\gamma}_1' \cap \partial\bar{\gamma})$ consists of a single point. Then we can slide $\bar{\gamma}$ ($\bar{\gamma}_1'$ resp.) to γ (γ' resp.) along the disk δ_γ ($\delta_{\gamma'}$ resp.). If $\bar{\gamma}_2' \cap \bar{\gamma} \neq \emptyset$, then $\bar{\gamma}_2' \cap \bar{\gamma} = \partial\bar{\gamma}_2' \cap \bar{\gamma}$ consists of a single point. Hence $\bar{\gamma}_1' \cup \bar{\gamma}$ composes an unknotted cycle and hence Lemma 4.1.8 holds (cf. Fig. 4.34). Otherwise, we can further isotope Σ to reduce (W_Σ, n_Σ), a contradiction (cf. Fig. 4.35).

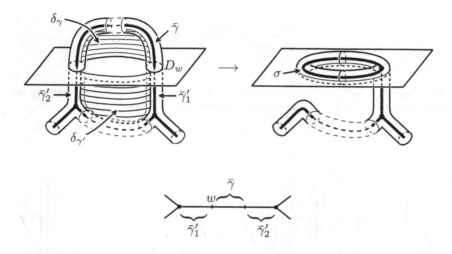

Fig. 4.34   Case B-(2)-(3). If $\operatorname{int}(\bar{\gamma}) \cap \operatorname{int}(\bar{\gamma}') = \emptyset$ and $\bar{\gamma}_2' \cap \bar{\gamma} \neq \emptyset$, we obtain an unknotted cycle from $\bar{\gamma}_1'$ and $\bar{\gamma}$.

We next suppose that $\operatorname{int}\bar{\gamma} \cap \operatorname{int}\bar{\gamma}' \neq \emptyset$. We may assume that $\operatorname{int}\bar{\gamma} \cap \operatorname{int}\bar{\gamma}_1' \neq \emptyset$.

Then there are two possibilities: $\operatorname{int}\bar{\gamma} \cap \operatorname{int}\bar{\gamma}_2' = \emptyset$ and $\operatorname{int}\bar{\gamma} \cap \operatorname{int}\bar{\gamma}_2' \neq \emptyset$. In each case, we can slide $\bar{\gamma}$ to γ along the disk δ_γ. Moreover, we can isotope σ and σ' slightly to reduce (W_Σ, n_Σ), a contradiction (cf. Fig. 4.36 and Fig. 4.37).

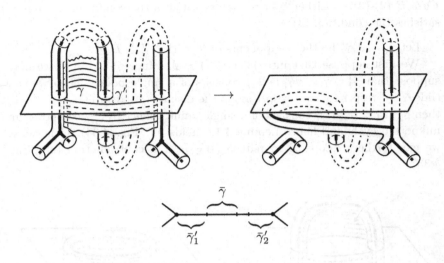

Fig. 4.35   Case B-(2)-(3). If $\mathrm{int}(\bar{\gamma}) \cap \mathrm{int}(\bar{\gamma}') = \emptyset$ and $\bar{\gamma}_2' \cap \bar{\gamma} = \emptyset$, the edges $\bar{\gamma}$ and $\bar{\gamma}_1'$ are isotoped along δ_γ and $\delta_{\gamma'}$ respectively so that (W_Σ, n_Σ) is reduced.

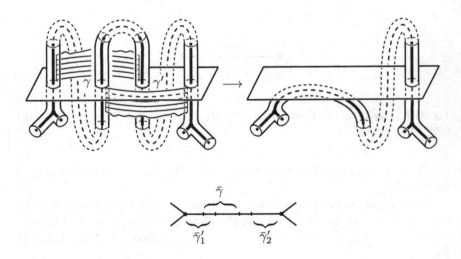

Fig. 4.36   Case B-(2)-(3). If $\mathrm{int}\bar{\gamma} \cap \mathrm{int}\bar{\gamma}' \neq \emptyset$ and $\mathrm{int}\bar{\gamma} \cap \mathrm{int}\bar{\gamma}_2' = \emptyset$, the edge $\bar{\gamma}$ is isotoped along δ_γ so that (W_Σ, n_Σ) is reduced.

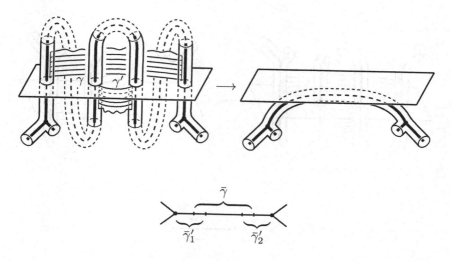

Fig. 4.37   Case B-(2)-(3). If $\operatorname{int}\bar{\gamma}\cap\operatorname{int}\bar{\gamma}' \neq \emptyset$ and $\operatorname{int}\bar{\gamma}\cap\operatorname{int}\bar{\gamma}_2' \neq \emptyset$, the edge $\bar{\gamma}$ is isotoped along δ_γ so that (W_Σ, n_Σ) is reduced.

Case B-(3)-(3).   Both $\bar{\gamma}$ and $\bar{\gamma}'$ satisfy the condition (3).

Let $\bar{\gamma}_1$ and $\bar{\gamma}_2$ ($\bar{\gamma}_1'$ and $\bar{\gamma}_2'$ resp.) be the components of $\bar{\gamma}$ ($\bar{\gamma}'$ resp.) with $h_{\bar{\gamma}_1}^1 \supset D_w$ ($h_{\bar{\gamma}_1'}^1 \supset D_{w'}$ resp.). Without loss of generality, we may suppose that $\bar{\gamma}_1 \subset \bar{\gamma}_1'$. Then there are teo possibilities: (1) $\bar{\gamma}_2 \subset \bar{\gamma}_2'$ and (2) $\bar{\gamma}_2 \supset \bar{\gamma}_2'$. In each case, we can slide $\bar{\gamma}_1'$ into γ' along the disk $\delta_{\gamma'}$. Moreover, we can isotope Σ to reduce (W_Σ, n_Σ) is reduced, a contradiction (cf. Fig. 4.38 and Fig. 4.39). $\qquad\square$

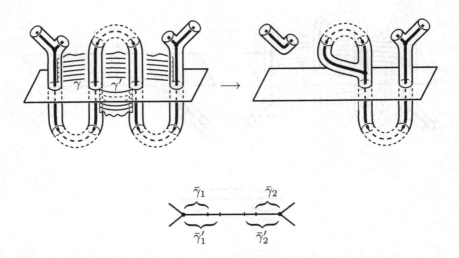

Fig. 4.38   Case B-(3)-(3). If $\bar{\gamma}_2 \subset \bar{\gamma}_2'$, the edge $\bar{\gamma}_1'$ is isotoped along $\delta_{\gamma'}$ so that (W_Σ, n_Σ) is reduced.

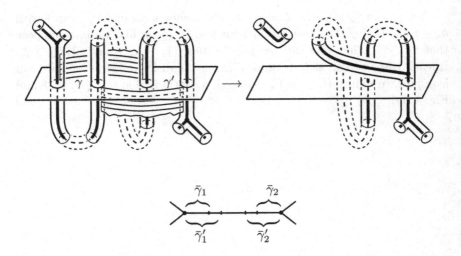

Fig. 4.39   Case B-(3)-(3). If $\bar{\gamma}_2 \supset \bar{\gamma}_2'$, the edge $\bar{\gamma}_1'$ is isotoped along $\delta_{\gamma'}$ so that (W_Σ, n_Σ) is reduced.

Let t_0^+ (t_0^- resp.) be the first critical height above t_0 (below t_0 resp.). Since $|P(t_0) \cap \Sigma| = W_\Sigma = \max\{w_\Sigma(t)|t \in (-1, 1)\}$, we see that the critical point of the height t_0^+ (t_0^- resp.) is a maximum or a λ-vertex (a minimum or a y-vertex resp.) (see Fig. 4.40).

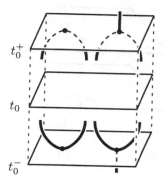

Fig. 4.40 The critical point of the height t_0^+ (t_0^- resp.) is a maximum or a λ-vertex (a minimum or a y-vertex resp.).

Lemma 4.1.9. *The critical height t_0^- is a y-vertex (not a minimum), or Σ is modified by edge slides so that the modified graph contains an unknotted cycle.*

Proof. Suppose that the critical point of the height t_0^- is a minimum. Let t_0^{-+} be a regular height just above t_0^-. Then $\Lambda(t_0^{-+})$ contains a fat-vertex with a lower simple outermost edge for the fat-vertex of $\Lambda(t_0^{-+})$. Hence it follows from Lemma 4.1.8 that every simple outermost edge for each fat-vertex of $\Lambda(t_0^{-+})$ is lower. Similarly, every simple outermost edge for each fat-vertex of $\Lambda(t_0)$ is upper. We now vary t for t_0^{-+} to t_0. Note that for each regular height t, all the simple outermost edges for each fat-vertex of $\Lambda(t)$ are either upper or lower (Lemma 4.1.8); such a regular height t is said to be upper or lower respectively. In these words, t_0^{-+} is lower and t_0 is upper.

Let $c_1, \ldots, c_n$ ($c_1 < \cdots < c_n$) be the critical heights of $h|_{\Delta_2}$ contained in $[t_0^{-+}, t_0]$. Note that the property 'upper' or 'lower' is unchanged at any height of $[t_0^{-+}, t_0] \setminus \{c_1, \ldots, c_n\}$. Hence there exists a critical height c_i such that a height t is changed from lower to upper at c_i. The graph $\Lambda(t)$ is changed as in Fig. 4.41 around the critical height c_i.

Let c_i^+ (c_i^- resp.) be a regular height just above (below resp.) c_i. We

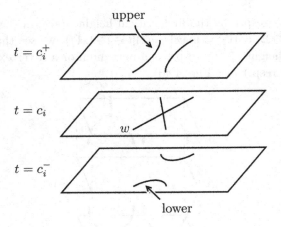

Fig. 4.41   Transformation of the graph $\Lambda(t)$ around the critical height c_i.

note that the lower disk for $\Lambda(c_i^-)$ and the upper disk for $\Lambda(c_i^+)$ in Fig. 4.41 are contained in the same component of Δ_2, say D. We take parallel copies, say D' and D'', of D such that D' is obtained by pushing D into one side and that D'' is obtained by pushing D into the other side (cf. Fig. 4.42). Then we may suppose that there is an upper (a lower resp.) simple outermost edge for a fat-vertex in D' (D'' resp.). Hence we can apply the arguments of the proof of Lemma 4.1.7 to modify Σ so that the modified graph contains an unknotted cycle. □

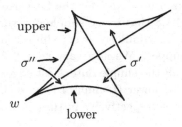

Fig. 4.42

Let v^- be the y-vertex of Σ at the height t_0^- and t_0^{--} a regular height just below t_0^-. Let v^{--} be the intersection point of the descending edges from v^- in Σ and $P(t_0^{--})$, and let $D_{v^{--}}$ be the fat-vertex of $\Lambda(t_0^{--})$ corresponding to v^{--}.

Lemma 4.1.10. *Every simple outermost edge for any fat-vertex of $\Lambda(t_0^{--})$ is lower, or Σ is modified by edge slides so that the modified graph contains an unknotted cycle.*

Proof. Suppose that there is a fat-vertex D_w of $\Lambda(t_0^{--})$ such that $\Lambda(t_0^{--})$ contains an upper simple outermost edge γ for D_w. Let σ be the edge of Σ with $h_\sigma^1 \supset D_w$. Let δ_γ be the outermost disk for (D_w, γ). Let $\bar\gamma$ ($\bar\gamma'$ resp.) be a union of the components obtained by cutting σ by the two fat-vertices of $\Lambda(t_0)$ incident to γ (γ' resp.) such that a 1-handle corresponding to each component intersects $\partial\delta_\gamma \setminus \gamma$ ($\partial\delta_{\gamma'} \setminus \gamma'$ resp.). Then $\bar\gamma$ ($\bar\gamma'$ resp.) satisfies one of the conditions (1), (2) and (3) in the proof of Lemma 4.1.7.

Case A. $D_{v^{--}} \neq D_w$.

Then we have the following three cases. In each case, we can slide (a component of) $\bar\gamma$ to γ along the disk δ_γ. Moreover, we can isotope Σ to reduce (W_Σ, n_Σ), a contradiction.

Case A-(1). $\bar\gamma$ satisfies the condition (1).

Then there are two possibilities: (i) $v^{--} \notin \bar\gamma$ and (ii) $v^{--} \in \bar\gamma$. In each case, see Fig. 4.43.

Case A-(2). $\bar\gamma$ satisfies the condition (2).

See Fig. 4.44.

Case A-(3). $\bar\gamma$ satisfies the condition (3).

Then there are two possibilities: (i) $v^{--} \notin \bar\gamma$ and (ii) $v^{--} \in \bar\gamma$. In each case, see Fig. 4.45.

Case B. $D_{v^{--}} = D_w$.

Since δ_γ is upper, we see that $\bar\gamma$ does not satisfy the condition (2). Hence we have the following.

Case B-(1). $\bar\gamma$ satisfies the condition (1).

Since γ is upper, we see that the y-vertex of Σ at the height t_0^- is an endpoint of $\bar\gamma$, *i.e.*, $\bar\gamma$ is the short vertical arc joining v^- to v^{--}. Then we can slide $\bar\gamma$ to γ along the disk δ_γ to obtain a new graph Σ'. Note that $(W_{\Sigma'}, n_{\Sigma'}) = (W_\Sigma, n_\Sigma)$ (cf. Fig. 4.46). However, the critical point for Σ' corresponding to v^- is a minimum. Hence we can apply the arguments in the proof of Lemma 4.1.9 to show that there is an unknotted cycle in Σ'.

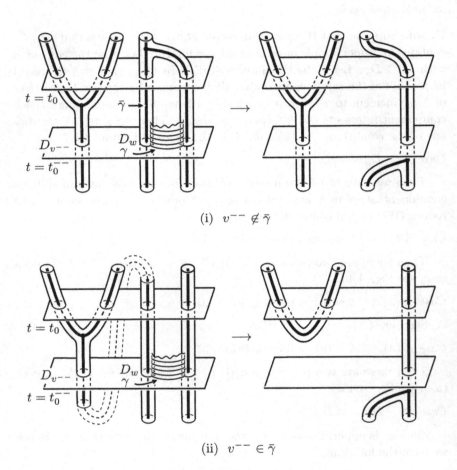

(i) $v^{--} \notin \bar{\gamma}$

(ii) $v^{--} \in \bar{\gamma}$

Fig. 4.43   Case A-(1).

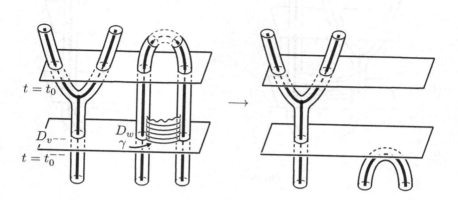

Fig. 4.44   Case A-(2).

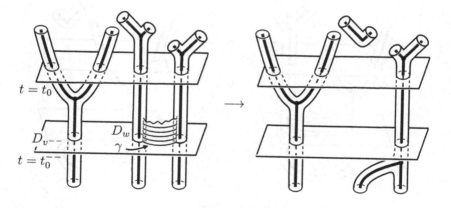

Fig. 4.45   Case A-(3).

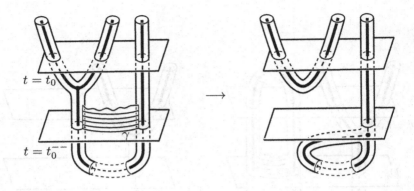

Fig. 4.46   Case B-(1).

Case B-(3). $\bar{\gamma}$ satisfies the condition (3).

Let $\bar{\gamma}_1$ and $\bar{\gamma}_2$ be the components of $\bar{\gamma}$ with $\partial\bar{\gamma}_1 \ni v^-$. Then we can slide $\bar{\gamma}_2$ into γ along the disk δ_γ. Moreover, we can isotope Σ to reduce (W_Σ, n_Σ), a contradiction (cf. Fig. 4.47). $\square$

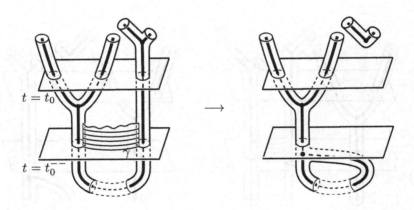

Fig. 4.47   Case B-(3).

Lemma 4.1.11. *Every simple outermost edge for any fat-vertex of* $\Lambda(t_0^{--})$ *is incident to* $D_{v^{--}}$, *or* Σ *is modified so that there is an unknotted cycle.*

Proof. Suppose that $\Lambda(t_0^{--})$ contains a simple outermost edge γ for D_w and is not incident to D_{v---}. Then it follows from Lemma 4.1.10 that γ is lower. This means that $\Lambda(t_0)$ contains a lower simple edge, because an edge disjoint from D_{v--} is not affected at all in $[t_0^{--}, t_0]$. This contradicts Lemma 4.1.8. □

We now prove Proposition 4.1.4.

Proof of Proposition 4.1.4. We first prove the following.

Claim. For any fat-vertex $D_w(\neq D_{v--})$ of $\Lambda(t_0^{--})$, there are no loops of $\Lambda(t_0^{--})$ based on D_w, or Σ is modified by edge slides so that the modified graph contains an unknotted cycle.

Proof. Suppose that there is a fat-vertex $D_w(\neq D_{v--})$ of $\Lambda(t_0^{--})$ such that there is a loop α of $\Lambda(t_0^{--})$ based on D_w. Then α separates $\mathrm{cl}(P(t_0^{--}) \setminus D_w)$ into two disks E_1 and E_2 with $D_{v--} \subset E_2$. By retaking D_w and α, if necessary, we may suppose that there are no loop components of $\Lambda(t_0^{--})$ in $\mathrm{int}(E_1)$. It follows from Lemma 4.1.5 that there is a fat-vertex $D_{w'}$ of $\Lambda(t_0^{--})$ in $\mathrm{int}(E_1)$. Then every outermost edge for $D_{w'}$ of $\Lambda(t_0^{--})$ is simple. Hence it follows from Lemma 4.1.11 that Σ contains an unknotted cycle and therefore we have the claim.

Then we have the following cases.

Case A. The descending edges of Σ from the maximum or λ-vertex v^+ at the height t_0^+ are equal to the ascending edges from v^- (cf. Fig. 4.48).

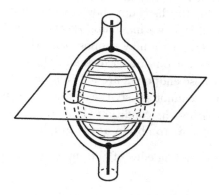

Fig. 4.48   Case A.

Then we can immediately see that there is an unknotted cycle.

Case B.   Exactly one of the descending edges from v^+ is equal to one of the ascending edges from v^- (cf. Fig. 4.49).

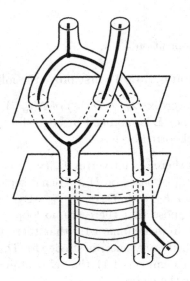

Fig. 4.49   Case B.

Let σ' be the other edge disjoint from v^-, and let w^{--} be the first intersection point of $P(t_0^{--})$ and the edge σ'. Let γ be an outermost edge for $D_{w^{--}}$ of $\Lambda(t_0^{--})$. By the claim above, we see that γ is simple. It follows from Lemma 4.1.10 that we may suppose that γ is lower. It also follows from Lemma 4.1.11 that we may suppose that the endpoints of γ are v^{--} and w^{--}. Let δ_γ be the outermost disk for (D_w, γ). Set $\bar{\gamma} = \sigma' \cap \delta_\gamma$. Since the subarc of σ' whose endpoints are v^{--} and w^{--} is monotonous and γ is lower, we see that $\bar{\gamma}$ cannot satisfy the condition (3) in the proof of Lemma 4.1.8. Hence $\bar{\gamma}$ satisfies the condition (1) or (2). In each case, we can slide $\bar{\gamma}$ to γ along the disk δ_γ to obtain a new graph with an unknotted cycle.

Case C.   Any descending edge of Σ from v^+ is disjoint from an ascending edge from v^-.

It follows from 4.1.10, 4.1.11 and the claim that $\Lambda(t^{--})$ contains a lower simple outermost edge γ_i $(i = 1, 2)$ for D_{w_i} which is incident to D_{w_i} and

D_{v--}. Let δ_{γ_i} be the outermost disk for (D_{w_i}, γ_i). Set $\bar{\gamma}_i = \sigma_i \cap \delta_{\gamma_i}$. Since the subarc of σ_i whose endpoints v^+ and w_i are monotonous and δ_{γ_i} is lower, we see that γ_i cannot satisfy the condition (3). Then we have the following.

Case C-(1). Both $\bar{\gamma}_1$ and $\bar{\gamma}_2$ satisfy the condition (1).

If $\sigma_1 = \sigma_2$, then we can slide $\bar{\gamma}_1$ to γ_1 along the disk δ_{γ_1}. We can further isotope Σ to reduce (W_Σ, n_Σ), a contradiction (cf. Fig. 4.50). Hence $\sigma_1 \neq \sigma_2$.

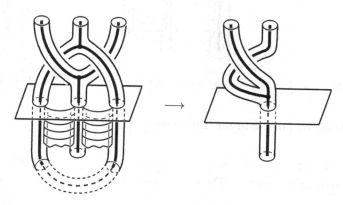

Fig. 4.50 Case C-(1). If $\sigma_1 = \sigma_2$, the edge $\bar{\gamma}_1$ is isotoped along δ_{γ_1} so that (W_Σ, n_Σ) is reduced.

Then we can slide $\bar{\gamma}_1 \cup \bar{\gamma}_2$ to $\gamma_1 \cup \gamma_2$ along $\delta_{\gamma_1} \cup \delta_{\gamma_2}$ so that a new graph contains an unknotted cycle (cf. Fig. 4.51).

Case C-(2). Either $\bar{\gamma}_1$ or $\bar{\gamma}_2$, say $\bar{\gamma}_1$, satisfies the condition (2).

Since $\bar{\gamma}_1$ satisfies the condition (2), we see that the endpoints of σ_1 are v^+ and v^-. Hence $w_2 \notin \sigma_1$. This implies that $\bar{\gamma}_2$ satisfies the condition (1). Then we first slide $\bar{\gamma}_2$ to γ_2 along δ_{γ_2}. We can further slide $\bar{\gamma}_1$ to γ_1 along δ_{γ_1} so that a new graph contains an unknotted cycle.

This completes the proof of Proposition 4.1.4. □

Remark 4.1.12. Any Heegaard splitting of a lens space is also obtained from a unique Heegaard splitting of genus one by stabilization. This was originally proved by Bonahon and Otal. See [Bonahon and Otal (1983)]. An alternative proof was given in [Rubinstein and Scharlemann (1996)].

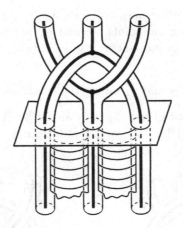

Fig. 4.51   Case C-(1). If $\sigma_1 \neq \sigma_2$, we can obtain an unknotted cycle from $\bar{\gamma}_1 \cup \bar{\gamma}_2$.

4.2   Reidemeister-Singer Theorem

Corollary 4.2.1. *Let M be a compact 3-manifold and $(C_1, C_2; S)$ a reducible Heegaard splitting. Then M is reducible or $(C_1, C_2; S)$ is stabilized.*

Proof. Suppose that M is irreducible. Let P be a 2-sphere such that $P \cap S$ is an essential simple closed curve. Since M is irreducible, we see that P bounds a 3-ball in M. Hence we can regard M as a connected sum of $\mathbb{S}^3$ and M. It follows from Theorem 4.1.1 that the induced Heegaard splitting of $\mathbb{S}^3$ is stabilized. Hence this cancelling pair of disks shows that $(C_1, C_2; S)$ is stabilized. $\square$

Corollary 4.2.2. *Any Heegaard splitting of a handlebody is standard, i.e, is obtained from a trivial splitting by stabilization.*

Exercise 4.2.3. Show Corollary 4.2.2.

Theorem 4.2.4 (Reidemeister-Singer Theorem). *Let M be a closed 3-manifold. Let $(C_1, C_2; S)$ and $(C_1', C_2'; S')$ be Heegaard splittings of M. Then there is a Heegaard splitting which is obtained by stabilization both of $(C_1, C_2; S)$ and of $(C_1', C_2'; S')$.*

Proof. Let Σ_{C_1} and $\Sigma_{C_1'}$ be spines of C_1 and C_1' respectively. By an isotopy, we may assume that $\Sigma_{C_1} \cap \Sigma_{C_1'} = \emptyset$ and $C_1 \cap C_1' = \emptyset$. Set $M' =$

$\text{cl}(M \setminus (C_1 \cup C_1'))$, $\partial_1 M' = \partial C_1$ and $\partial_2 M' = \partial C_2$. Let $(\bar{C}_1, \bar{C}_2; \bar{S})$ be a Heegaard splitting of $(M'; \partial_1 M', \partial_2 M')$. Set $C_1^* = C_1 \cup \bar{C}_1$ and $C_2^* = C_2 \cup \bar{C}_2$. Then it is easy to see that $(C_1^*, C_2^*; \bar{S})$ is a Heegaard splitting of M. Note that $C_2' = C_1 \cup M' = C_1 \cup (\bar{C}_1 \cup \bar{C}_2) = (C_1 \cup \bar{C}_1) \cup \bar{C}_2$. Here, we note that $(C_1^*, \bar{C}_2; \bar{S})$ is a Heegaard splitting of C_2'. It follows from Corollary 4.2.2 that $(C_1^*, \bar{C}_2; \bar{S})$ is obtained from a trivial splitting of C_2' by stabilization. This implies that $(C_1^*, C_2^*; \bar{S})$ is obtained from $(C_1', C_2'; S')$ by stabilization. On the argument above, by replacing C_1 to C_1', we see that $(C_1^*, C_2^*; \bar{S})$ is also obtained from $(C_1, C_2; S)$ by stabilization. □

Remark 4.2.5. The stabilization problem is one of the most important themes on Heegaard theory. But we do not give any more here. For the detail, for example, see [Lei (2000)], [Reidemeister (1933)], [Rubinstein and Scharlemann (1996)], [Sedgwick (1997)] and [Singer (1933)]. We will also give some comments on the stabilization problem at Section 5.5.

Chapter 5

Generalized Heegaard splittings

In 1992, Scharlemann and Thompson defined the width of 3-manifolds and showed that manifold decompositions minimizing this width possessed many valuable properties. (See [Scharlemann and Thompson (1994b)].) The key insight is captured in Proposition 5.2.3. It tells us that a thin manifold decomposition provides a sequence of strongly irreducible Heegaard splittings meeting along incompressible surfaces. The notion thus refines the weak reductions inspired by the insights in [Casson and Gordon (1987)]. See also [Kobayashi (2003)].

Working with thin generalized Heegaard splittings requires careful bookkeeping. Fork complexes are envisioned as a tool that accounts for the boundary components of the compression bodies in a generalized Heegaard splitting.

5.1 Definitions

Definition 5.1.1. A *0-fork* is a connected oriented 1-complex obtained by joining a point p to a point g with a 1-simplex oriented toward g and away from p. For $n \geq 1$, an *n-fork* is a connected oriented 1-complex obtained by joining a point p to each of n distinct points t_i $(i = 1, ..., n)$ and to a point g such that the 1-simplices are oriented toward g and away from t_i. We call p the *root*, each t_i a *tine* and g the *grip* of the n-fork.

Remark 5.1.2. In what follows, n-forks will be used to account for the boundary components of compression bodies. For a compression body C, the t_i $(i = 1, 2, ..., n)$ correspond to the components of $\partial_- C$ and g corresponds to $\partial_+ C$ (cf. Fig. 5.1).

Definition 5.1.3. Let $\mathcal{A}$ ($\mathcal{B}$ resp.) be a collection of finite forks, $T_{\mathcal{A}}$ ($T_{\mathcal{B}}$

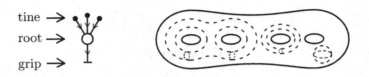

Fig. 5.1   An n-fork corresponds to a compression body.

resp.) a collection of tines of $\mathcal{A}$ ($\mathcal{B}$ resp.) and $G_{\mathcal{A}}$ ($G_{\mathcal{B}}$ resp.) a collection of grips of $\mathcal{A}$ ($\mathcal{B}$ resp.). We suppose that there are bijections $\mathcal{T} : T_{\mathcal{A}} \to T_{\mathcal{B}}$ and $\mathcal{G} : G_{\mathcal{A}} \to G_{\mathcal{B}}$. A *fork complex* $\mathscr{F}$ is a connected oriented 1-complex $\mathcal{A} \cup (-\mathcal{B}) / \{\mathcal{T}, \mathcal{G}\}$, where $-\mathcal{B}$ denotes the 1-complex obtained by taking the opposite orientation of each 1-simplex and the equivalence relation $/\{\mathcal{T}, \mathcal{G}\}$ is given by $t \sim \mathcal{T}(t)$ for any $t \in T_{\mathcal{A}}$ and $g \sim \mathcal{G}(g)$ for any $g \in G_{\mathcal{A}}$. We define

$$\partial_1 \mathscr{F} = \{(\text{tines of } \mathcal{A}) \setminus T_{\mathcal{A}}\} \cup \{(\text{grips of } \mathcal{B}) \setminus G_{\mathcal{B}}\}$$

and

$$\partial_2 \mathscr{F} = \{(\text{tines of } \mathcal{B}) \setminus T_{\mathcal{B}}\} \cup \{(\text{grips of } \mathcal{A}) \setminus G_{\mathcal{A}}\}.$$

Definition 5.1.4. A fork complex is *exact* if there exists $e \in \mathrm{Hom}(C_0(\mathscr{F}), \mathbb{R})$ such that

(1) $e(v_1) = 0$ for any $v_1 \in \partial_1 \mathscr{F}$,
(2) $(\delta e)(e_{\mathcal{A}}) > 0$ for any 1-simplex $e_{\mathcal{A}}$ in $\mathcal{A}$ with the standard orientation, $(\delta e)(e_{\mathcal{B}}) < 0$ for any 1-simplex $e_{\mathcal{B}}$ in $\mathcal{B}$ with the standard orientation, where δ denotes the coboundary operator $\mathrm{Hom}(C_0(\mathscr{F}), \mathbb{R}) \to \mathrm{Hom}(C_1(\mathscr{F}), \mathbb{R})$, and
(3) $e(v_2) = 1$ for any $v_2 \in \partial_2 \mathscr{F}$.

Remark 5.1.5. Geometrically, $\mathscr{F}$ is exact if and only if we can put $\mathscr{F}$ in $\mathbb{R}^3$ so that

(1) $\partial_1 \mathscr{F}$ lies in the plane of height 0,
(2) for any path α in $\mathscr{F}$ from a point in $\partial_1 \mathscr{F}$ to a point in $\partial_2 \mathscr{F}$, $h|_\alpha$ is monotonically increasing, where h is the height function of $\mathbb{R}^3$, and
(3) $\partial_2 \mathscr{F}$ lies in the plane of height 1 (cf. Fig. 5.2).

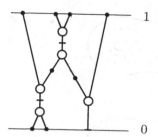

Fig. 5.2 An example of an exact fork complex.

In the following, we regard fork complexes as geometric objects, *i.e.*, 1-dimensional polyhedra.

Definition 5.1.6. A *fork of $\mathscr{F}$* is the image (in $\mathscr{F}$) of a fork in $\mathcal{A} \cup \mathcal{B}$. A *grip* (a *root* and a *tine* resp.) *of $\mathscr{F}$* is the image (in $\mathscr{F}$) of a grip (a root and a tine resp.) in $\mathcal{A} \cup \mathcal{B}$.

Definition 5.1.7. Let M be a compact orientable 3-manifold, and let $(\partial_1 M, \partial_2 M)$ be a partition of boundary components of M. A *generalized Heegaard splitting* of $(M; \partial_1 M, \partial_2 M)$ is a pair of an exact fork complex $\mathscr{F}$ and a proper map $\rho : (M; \partial_1 M, \partial_2 M) \to (\mathscr{F}; \partial_1 \mathscr{F}, \partial_2 \mathscr{F})$ which satisfies the following.

(1) The map ρ is transverse to $\mathscr{F} - \{\text{the roots of } \mathscr{F}\}$.
(2) For each fork $\mathcal{F} \subset \mathscr{F}$, we have the following (cf. Fig. 5.3).
   (a) If $\mathcal{F}$ is a 0-fork, then $\rho^{-1}(\mathcal{F})$ is a handlebody $V_{\mathcal{F}}$ such that (i) $\rho^{-1}(g) = \partial V_{\mathcal{F}}$ and (ii) $\rho^{-1}(p)$ is a 1-complex which is a spine of $V_{\mathcal{F}}$, where g is the grip of $\mathcal{F}$.
   (b) If $\mathcal{F}$ is an n-fork with $n \geq 1$, then $\rho^{-1}(\mathcal{F})$ is a connected compression body $V_{\mathcal{F}}$ such that (i) $\rho^{-1}(g) = \partial_+ V_{\mathcal{F}}$, (ii) for each tine t_i, $\rho^{-1}(t_i)$ is a connected component of $\partial_- V_{\mathcal{F}}$ and $\rho^{-1}(t_i) \neq \rho^{-1}(t_j)$ for $i \neq j$ and (iii) $\rho^{-1}(p)$ is a 1-complex which is a deformation retract of $V_{\mathcal{F}}$, where g is the grip of $\mathcal{F}$, p is the root of $\mathcal{F}$ and $\{t_i\}_{1 \leq i \leq n}$ is the set of the tines of $\mathcal{F}$.

Remark 5.1.8. Let g be a grip of $\mathscr{F}$ which is contained in the interior of $\mathscr{F}$. Let $\mathcal{F}_1$ and $\mathcal{F}_2$ be the forks of $\mathscr{F}$ which are adjacent to g. Then $(\rho^{-1}(\mathcal{F}_1), \rho^{-1}(\mathcal{F}_2); \rho^{-1}(g))$ is a Heegaard splitting of $\rho^{-1}(\mathcal{F}_1 \cup \mathcal{F}_2)$.

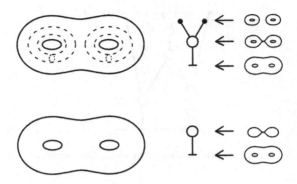

Fig. 5.3

Definition 5.1.9. A generalized Heegaard splitting $(\mathscr{F}, \rho)$ is said to be *strongly irreducible* if (1) for each tine t, $\rho^{-1}(t)$ is incompressible, and (2) for each grip g with two forks attached to g, say $\mathcal{F}_1$ and $\mathcal{F}_2$, $(\rho^{-1}(\mathcal{F}_1), \rho^{-1}(\mathcal{F}_2); \rho^{-1}(g))$ is strongly irreducible.

Let $\mathcal{M}$ be the set of finite multisets of $\mathbb{Z}_{\geq 0} = \{0, 1, 2, ...\}$. We define a total order $<$ on $\mathcal{M}$ as follows. For M_1 and $M_2 \in \mathcal{M}$, we first arrange the elements of M_i ($i = 1, 2$) in non-increasing order respectively. Then we compare the arranged tuples of non-negative integers by lexicographic order.

Example 5.1.10. (1) If $M_1 = \{5, 4, 1, 1, 1\}$ and $M_2 = \{5, 3, 2, 2, 2, 1\}$, then $M_2 < M_1$.
(2) If $M_1 = \{3, 1, 0, 0\}$ and $M_2 = \{3, 1, 0, 0, 0\}$, then $M_1 < M_2$.

Definition 5.1.11. Let $(\mathscr{F}, \rho)$ be a generalized Heegaard splitting of $(M; \partial_1 M, \partial_2 M)$. We define *the width of* $(\mathscr{F}, \rho)$ to be the multiset

$$w(\mathscr{F}, \rho) = \{\text{genus}(\rho^{-1}(g_1)), \ldots, \text{genus}(\rho^{-1}(g_m))\},$$

where $\{g_1, \ldots, g_m\}$ is the set of the grips of $\mathscr{F}$. We say that $(\mathscr{F}, \rho)$ is *thin* if $w(\mathscr{F}, \rho)$ is minimal among all generalized Heegaard splittings of $(M; \partial_1 M, \partial_2 M)$.

Example 5.1.12. A thin generalized Heegaard splitting of the 3-ball B^3 is one of the two fork complexes illustrated in Fig. 4.4, where $\rho^{-1}(\mathcal{F}_1)$ is a 3-ball and $\rho^{-1}(\mathcal{F}_2) \cong \mathbb{S}^2 \times [0, 1]$.

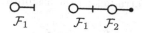

Fig. 5.4   The thin generalized Heegaard splittings of the 3-ball.

5.2   Thin generalized Heegaard splittings

Throughout this section, we let $(\mathscr{F}, \rho)$ denote a thin generalized Heegaard splitting of the triple $(M; \partial_1 M, \partial_2 M)$.

Observation 5.2.1. Let t be a tine of $\mathscr{F}$. Then any 2-sphere component of $\rho^{-1}(t)$ is essential in M unless M is a 3-ball.

Proof. Suppose that there is a tine t such that $\rho^{-1}(t)$ is a 2-sphere, say P, which bounds a 3-ball B in M. Let $\mathscr{F}_B$ be the subcomplex of $\mathscr{F}$ with $\rho^{-1}(\mathscr{F}_B) = B$. If $\mathscr{F}_B = \mathscr{F}$, then we see that M is a 3-ball. Otherwise, there is a fork $\mathcal{F}'$ with $t \in F'$ and $F' \not\subset \mathscr{F}_B$. Let e_t be the 1-simplex in $\mathcal{F}'$ joining t to the root of $\mathcal{F}'$. Set $\mathscr{F}^* = \mathscr{F} \backslash (\mathscr{F}_B \cup e_t)$. Note that $\rho^{-1}(F' \cup \mathscr{F}_B)$ $(= \rho^{-1}(\mathcal{F}') \cup B)$ is a compression body V^*. Then it is easy to see that we can modify ρ in V^* to obtain $\rho^* : M \to \mathscr{F}^*$ such that $(\rho^*)^{-1}(\mathcal{F}' \backslash e_t)$ is the compression body V^* (cf. Fig. 5.5).

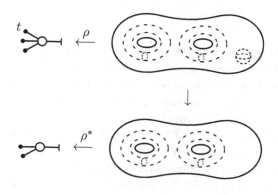

Fig. 5.5

Moreover, the generalized Heegaard structure on $(\mathscr{F}, \rho)$ (e.g. $\mathcal{A}, \mathcal{B}$ decomposition etc.) is naturally inherited to $(\mathscr{F}^*, \rho^*)$. Then we clearly have

$w(\mathscr{F}^*, \rho^*) < w(\mathscr{F}, \rho)$, contradicting the assumption that $(\mathscr{F}, \rho)$ is thin. $\square$

Lemma 5.2.2. *Suppose that there is a fork $\mathcal{F}$ such that $\rho^{-1}(\mathcal{F})$ is trivial. Let t be the tine of $\mathcal{F}$. Then $\rho^{-1}(\sqcup)$ is a component of ∂M and one of the following holds.*

(1) M is a (possibly punctured) 3-ball,
(2) $M \cong \rho^{-1}(t) \times [0,1]$, or
(3) $\rho^{-1}(t)$ is compressible in M.

Proof. We first prove that $\rho^{-1}(t)$ is a boundary component of M. Suppose, for a contradiction, that $\rho^{-1}(t)$ is not a boundary component of M. Let g be the grip of $\mathcal{F}$. If $\rho^{-1}(g)$ is a boundary component of M, then we can reduce the width by removing $\mathcal{F}$, a contradiction (cf. Fig. 5.6).

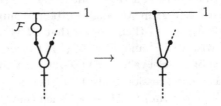

Fig. 5.6 If $\rho^{-1}(g)$ is a boundary component of M, then we can reduce the width by removing $\mathcal{F}$.

Hence $\mathcal{F}$ is contained in the interior of $\mathscr{F}$. Note that $\mathcal{F}$ is a 1-fork. Let $\mathcal{F}_1$ be the fork attached to g and $\mathcal{F}_2$ the fork attached to t. Note that since $\rho^{-1}(\mathcal{F})$ is a trivial compression body, we see that $\rho^{-1}(\mathcal{F}_1 \cup \mathcal{F} \cup \mathcal{F}_2)$ is also a compression body. Hence we can replace $\mathcal{F}_1 \cup \mathcal{F} \cup \mathcal{F}_2$ in $\mathscr{F}$ to a new fork so that we obtain a new fork complex, say $\mathscr{F}^*$. Moreover, we can modify $\rho : M \to \mathscr{F}$ to obtain $\rho^* : M \to \mathscr{F}^*$ so that $(\mathscr{F}^*, \rho^*)$ is a generalized Heegaard splitting of $(M; \partial_1 M, \partial_2 M)$ with $w(\mathscr{F}^*, \rho^*) < w(\mathscr{F}, \rho)$, a contradiction (cf. Fig. 5.7). Hence $\rho^{-1}(t)$ is a boundary component of M.

We next show that one of the conclusions (1)-(3) of Lemma 5.2.2 holds. Suppose that both conclusions (1) and (2) of Lemma 5.2.2 do not hold, *i.e.*, M is not a 3-ball and $M \not\cong \rho^{-1}(t) \times [0,1]$. Then there is a fork $\mathcal{F}'(\neq \mathcal{F})$ attached to g. Moreover, since $(\mathscr{F}, \rho)$ is thin and $M \not\cong \rho^{-1}(t) \times [0,1]$, we see that $\rho^{-1}(\mathcal{F}')$ is a non-trivial compression body. Also, since M is not a 3-ball, $\rho^{-1}(t)$ is not a 2-sphere. Hence we see that $\rho^{-1}(t)$ is compressible

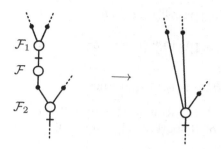

Fig. 5.7   We can replace $\mathcal{F}_1 \cup \mathcal{F} \cup \mathcal{F}_2$ in $\mathscr{F}$ to a new fork.

in $\rho^{-1}(\mathcal{F}) \cup \rho^{-1}(\mathcal{F}')$. This implies that the conclusion (3) of Lemma 5.2.2 holds.
□

Proposition 5.2.3 establishes that thin generalized Heegaard splittings are strongly irreducible. The converse is not true. This was first observed by Kobayashi, see [Kobayashi (2003)].

Proposition 5.2.3. *Let $\mathcal{F}_1$ and $\mathcal{F}_2$ be forks of $\mathscr{F}$ which have the same grip g of $\mathscr{F}$. Then $(\rho^{-1}(\mathcal{F}_1), \rho^{-1}(\mathcal{F}_2); \rho^{-1}(g))$ is strongly irreducible.*

Proof. Set $A_g = \rho^{-1}(\mathcal{F}_1)$, $B_g = \rho^{-1}(\mathcal{F}_2)$, $S_g = \rho^{-1}(g)$, $M_g = A_g \cup B_g$, $\partial_1 M_g = \partial_- A_g$ and $\partial_2 M_g = \partial_- B_g$. Then $(A_g, B_g; S_g)$ is a Heegaard splitting of $(M_g; \partial_1 M_g, \partial_2 M_g)$. Suppose that $(A_g, B_g; S_g)$ is weakly reducible. Let D_A and D_B be meridian disks of A_g and B_g respectively which satisfy $\partial D_A \cap \partial D_B = \emptyset$. Let Δ_A (Δ_B resp.) be a complete meridian system of A_g (B_g resp.) such that D_A (D_B resp.) is a component of Δ_A (Δ_B resp.) (cf. (6) of Remark 2.1.4). Note that A_g is obtained from $\partial_- A_g \times [0,1]$ and 0-handles $\mathcal{H}^0$ by attaching 1-handles $\mathcal{H}^1$ corresponding to Δ_A (cf. (3) of Remark 2.1.4) and that B_g is obtained from $S_g \times [0,1]$ by attaching 2-handles $\mathcal{H}^2$ corresponding to Δ_B and 3-handles $\mathcal{H}^3$ (cf. Definition 2.1.2). Hence we see that M_g admits the following decomposition (cf. Remark 2.1.14):

$$M_g = (\partial_1 M_g \times [0,1]) \cup \mathcal{H}^0 \cup \mathcal{H}^1 \cup \mathcal{H}^2 \cup \mathcal{H}^3.$$

Let h^1 be the component of $\mathcal{H}^1$ corresponding to D_A and h^2 the component of $\mathcal{H}^2$ corresponding to D_B. Then M_g admits the following decomposition:

$$M_g = (\partial_1 M_g \times [0,1]) \cup \mathcal{H}^0 \cup (\mathcal{H}^1 \setminus h^1) \cup h^2 \cup h^1 \cup (\mathcal{H}^2 \setminus h^2) \cup \mathcal{H}^3.$$

Set $A'_g = (\partial_1 M_g \times [0,1]) \cup \mathcal{H}^0 \cup (\mathcal{H}^1 \setminus h^1)$. We divide the proof into the following two cases.

Case 1. ∂D_A or ∂D_B is non-separating in S_g.

Suppose first that ∂D_A is non-separating in S_g. Then A'_g is a compression body (cf. (6) of Remark 2.1.4). Since $A'_g = (\partial_1 M_g \times [0,1]) \cup \mathcal{H}^0 \cup (\mathcal{H}^1 \setminus h^1)$, we obtain

$$M_g = A'_g \cup h^2 \cup h^1 \cup (\mathcal{H}^2 \setminus h^2) \cup \mathcal{H}^3.$$

Note that the attaching region of the 2-handle h^2 is contained in $\partial_+ A'_g$. Hence we have

$$M_g \cong A'_g \cup ((\partial_+ A'_g \times [0,1]) \cup h^2) \cup h^1 \cup (\mathcal{H}^2 \setminus h^2) \cup \mathcal{H}^3.$$

Set $B'_g = (\partial_+ A'_g \times [0,1]) \cup h^2$. Then B'_g is also a compression body and we have

$$M_g \cong A'_g \cup B'_g \cup h^1 \cup (\mathcal{H}^2 \setminus h^2) \cup \mathcal{H}^3.$$

Note that $\partial_- B'_g$ is homeomorphic to the surface obtained from S_g by performing surgery along $\partial D_A \cup \partial D_B$. Then we have the following subcases.

Case 1.1. ∂D_B is non-separating in S_g and $\partial D_A \cup \partial D_B$ is non-separating in S_g.

Then $\partial_- B'_g$ is connected. Note that

$$\begin{aligned} M_g &\cong A'_g \cup B'_g \cup h^1 \cup (\mathcal{H}^2 \setminus h^2) \cup \mathcal{H}^3 \\ &\cong A'_g \cup B'_g \cup ((\partial_- B'_g \times [0,1]) \cup h^1) \cup (\mathcal{H}^2 \setminus h^2) \cup \mathcal{H}^3. \end{aligned}$$

Set $A''_g = (\partial_- B'_g \times [0,1]) \cup h^1$. Then A''_g is also a compression body and we have

$$\begin{aligned} M_g &\cong A'_g \cup B'_g \cup A''_g \cup (\mathcal{H}^2 \setminus h^2) \cup \mathcal{H}^3 \\ &\cong A'_g \cup B'_g \cup A''_g \cup ((\partial_+ A''_g \times [0,1] \cup (\mathcal{H}^2 \setminus h^2) \cup \mathcal{H}^3). \end{aligned}$$

Set $B''_g = \partial_+ A''_g \times [0,1] \cup (\mathcal{H}^2 \setminus h^2) \cup \mathcal{H}^3$. Note that $B'_g \cap A''_g = \partial_- B'_g = \partial_- A''_g$. This shows that each handle of $\mathcal{H}^2 \setminus h^2$ and $\mathcal{H}^3$ is adjacent to A''_g along $\partial_+ A''_g$. This implies that B''_g is also a compression body. Hence we have

$$M_g \cong (A'_g \cup B'_g) \cup (A''_g \cup B''_g).$$

Then we can substitute $\mathcal{F}_1 \cup \mathcal{F}_2$ in $\mathcal{F}$ for $\mathcal{F}'_1 \cup \mathcal{F}'_2 \cup \mathcal{F}''_1 \cup \mathcal{F}''_2$, where $\mathcal{F}'_1$, $\mathcal{F}'_2$, $\mathcal{F}''_1$ and $\mathcal{F}''_2$ are forks corresponding to A'_g, B'_g, A''_g and B''_g respectively. Set $\mathcal{F}^* = (\mathcal{F} \setminus (\mathcal{F}_1 \cup \mathcal{F}_2)) \cup (\mathcal{F}'_1 \cup \mathcal{F}'_2 \cup \mathcal{F}''_1 \cup \mathcal{F}''_2)$. Then we can modify $\rho : M \to \mathcal{F}$ in M_g to obtain $\rho^* : M \to \mathcal{F}^*$ such that $(\rho^*)^{-1}(\mathcal{F}'_1) = A'_g$, $(\rho^*)^{-1}(\mathcal{F}'_2) = B'_g$, $(\rho^*)^{-1}(\mathcal{F}''_1) = A''_g$ and $(\rho^*)^{-1}(\mathcal{F}''_2) = B''_g$. It is easy to see that $w(\mathcal{F}^*, \rho^*) < w(\mathcal{F}, \rho)$, a contradiction (cf. Fig. 5.8).

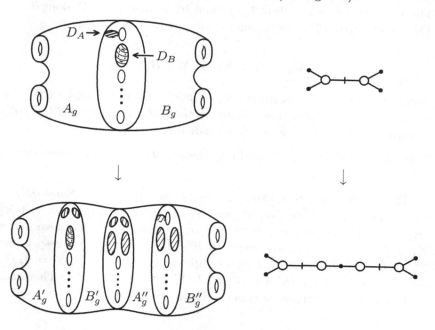

Fig. 5.8 Case 1.1. ∂D_B is non-separating in S_g and $\partial D_A \cup \partial D_B$ is non-separating in S_g.

Case 1.2. ∂D_B is non-separating in S_g and $\partial D_A \cup \partial D_B$ is separating in S_g.

Then $\partial_- B'_g$ consists of two components, say G_1 and G_2.

$$M_g \cong A'_g \cup B'_g \cup h^1 \cup (\mathcal{H}^2 \setminus h^2) \cup \mathcal{H}^3$$
$$\cong A'_g \cup B'_g \cup (\partial_- B'_g \times [0,1]) \cup h^1 \cup (\mathcal{H}^2 \setminus h^2) \cup \mathcal{H}^3$$
$$\cong A'_g \cup B'_g \cup ((G_1 \cup G_2) \times [0,1]) \cup h^1 \cup (\mathcal{H}^2 \setminus h^2) \cup \mathcal{H}^3.$$

Set $A''_g = ((G_1 \cup G_2) \times [0,1]) \cup h^1$. Since ∂D_B is non-separating in S_g, we see that h^1 joins G_1 to G_2. Hence A''_g is a compression body and we have

$$M_g \cong A'_g \cup B'_g \cup A''_g \cup (\mathcal{H}^2 \setminus h^2) \cup \mathcal{H}^3$$
$$\cong A'_g \cup B'_g \cup A''_g \cup (\partial_+ A''_g \times [0,1]) \cup (\mathcal{H}^2 \setminus h^2) \cup \mathcal{H}^3.$$

Set $B''_g = \partial_+ A''_g \times [0,1] \cup (\mathcal{H}^2 \setminus h^2) \cup \mathcal{H}^3$. Note that $B'_g \cap A''_g = \partial_- B'_g = \partial_- A''_g$. This shows that each handle of $\mathcal{H}^2 \setminus h^2$ and $\mathcal{H}^3$ is adjacent to A''_g along $\partial_+ A''_g$. This implies that B''_g is also a compression body. Hence we have

$$M_g \cong (A'_g \cup B'_g) \cup (A''_g \cup B''_g).$$

According to this decomposition, we can modify the fork complex $(\mathscr{F}, \rho)$ as in Fig. 5.9 or Fig. 5.10. It is easy to see that for a new complex $(\mathscr{F}^*, \rho^*)$, we have $w(\mathscr{F}^*, \rho^*) < w(\mathscr{F}, \rho)$, a contradiction.

Case 1.3. ∂D_B is separating in S_g (hence $\partial D_A \cup \partial D_B$ is separating in S_g).

Then $\partial_- B'_g$ consists of two components, say $\bar{G}_1$ and $\bar{G}_2$. Since ∂D_B is separating in S_g, we see that h^1 joins $\bar{G}_1$ or $\bar{G}_2$, say $\bar{G}_1$, to itself. Let $\mathcal{H}_1^2$ ($\mathcal{H}_2^2$ resp.) be the components of $\mathcal{H}^2 \setminus h^2$ adjacent to $\bar{G}_1$ ($\bar{G}_2$ resp.). Let $\mathcal{H}_1^3$ ($\mathcal{H}_2^3$ resp.) be the components of $\mathcal{H}^3$ adjacent to $\bar{G}_1$ ($\bar{G}_2$ resp.). Set $\bar{B}'_g = B'_g \cup \mathcal{H}_2^2 \cup \mathcal{H}_2^3$. Then $\bar{B}'_g$ is a compression body with $\partial_+ \bar{B}'_g = \partial_+ B'_g$. Set $A''_g = (\bar{G}_1 \times [0,1]) \cup h^1$ and $B''_g = (\partial_+ A''_g \times [0,1]) \cup \mathcal{H}_1^2 \cup \mathcal{H}_1^3$. Then each of A''_g and B''_g is a compression body. Hence we have

$$M_g \cong A'_g \cup B'_g \cup h^1 \cup (\mathcal{H}_1^2 \cup \mathcal{H}_2^2) \cup (\mathcal{H}_1^3 \cup \mathcal{H}_2^3)$$
$$\cong A'_g \cup (B'_g \cup \mathcal{H}_2^2 \cup \mathcal{H}_2^3) \cup h^1 \cup \mathcal{H}_1^2 \cup \mathcal{H}_1^3$$
$$\cong A'_g \cup \bar{B}'_g \cup (\bar{G}_1 \times [0,1]) \cup h^1 \cup \mathcal{H}_1^2 \cup \mathcal{H}_1^3$$
$$\cong A'_g \cup \bar{B}'_g \cup A''_g \cup (\partial_+ A''_g \times [0,1]) \cup \mathcal{H}_1^2 \cup \mathcal{H}_1^3$$
$$\cong (A'_g \cup \bar{B}'_g) \cup (A''_g \cup B''_g).$$

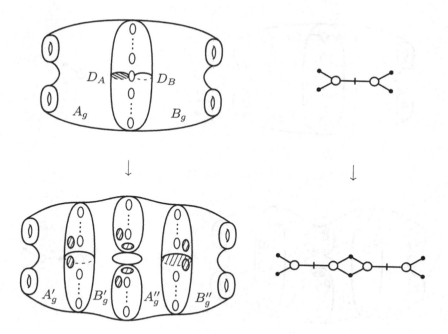

Fig. 5.9 Case 1.2. The case of irreducible splittings such that ∂D_B is non-separating in S_g and $\partial D_A \cup \partial D_B$ is separating in S_g.

According to this decomposition, we can modify the fork complex $(\mathscr{F}, \rho)$ as in Fig. 5.11. It is easy to see that for a new complex $(\mathscr{F}^*, \rho^*)$, we have $w(\mathscr{F}^*, \rho^*) < w(\mathscr{F}, \rho)$, a contradiction. Therefore if ∂D_A is non-separating, we have the desired conclusion.

Suppose next that ∂D_B is non-separating in S_g. Then we start with the dual handle decomposition:

$$M_g = (\partial_2 M_g \times [0,1]) \cup \bar{\mathcal{H}}^0 \cup \bar{\mathcal{H}}^1 \cup \bar{\mathcal{H}}^2 \cup \bar{\mathcal{H}}^3$$

and apply the above arguments which gives a contradiction.

Case 2.   Each of ∂D_A and ∂D_B is separating in S.

Then A'_g consists of two compression bodies, say $\bar{A}'_g$ and $\tilde{A}'_g$ (cf. (6) of Remark 2.1.4). We may suppose that h^2 is attached to $\partial_+ \bar{A}'_g$. Set $B'_g = (\partial_+ \bar{A}'_g \times [0,1]) \cup h^2$. Since D_B is separating in S, we see that $\partial_- B'_g$ consists of two components, say G_1 and G_2. Note that D_A is also separating

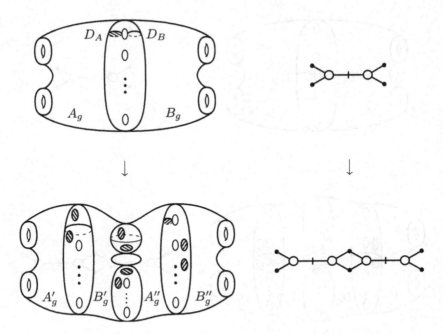

Fig. 5.10　Case 1.2. The case of reducible splittings such that ∂D_B is non-separating in S_g and $\partial D_A \cup \partial D_B$ is separating in S_g.

in S_g. Hence we may suppose that $h^1 \cap G_2 \neq \emptyset$ and $h^1 \cap G_1 = \emptyset$. Let $\mathcal{H}_1^2$ be the components of $\mathcal{H}^2 \setminus h^2$ adjacent to G_1 and $\mathcal{H}_1^3$ be the components of $\mathcal{H}^3$ adjacent to G_1. Set $\mathcal{H}_2^2 = \mathcal{H}^2 \setminus (h^2 \cup \mathcal{H}_1^2)$, $\mathcal{H}_2^3 = \mathcal{H}^3 \setminus \mathcal{H}_1^3$ and $\bar{B}_g' = B_g' \cup \mathcal{H}_1^2 \cup \mathcal{H}_1^3$. Then $\bar{B}_g'$ is a compression body. Set $A_g^* = (G_2 \times [0, 1]) \cup \tilde{A}_g' \cup h_1$ and $B_g'' = (\partial_+ A_g'' \times [0, 1]) \cup \mathcal{H}_2^2 \cup \mathcal{H}_2^3$. Note that each of A_g'' and B_g'' is a compression body. Set $A_g'' = \tilde{A}_g' \cup A_g^*$. Note also that A_g'' is a compression body (cf. (5) of Remark 2.1.4). Hence we have

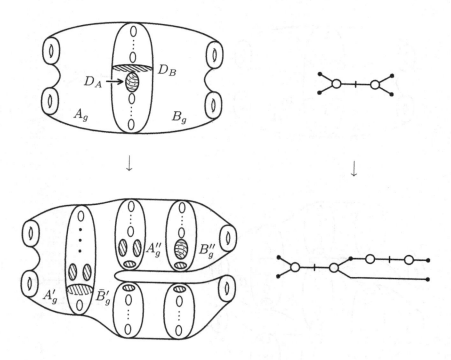

Fig. 5.11 Case 1.3. ∂D_B is separating in S_g.

$$M_g \cong A'_g \cup h^2 \cup h^1 \cup (\mathcal{H}_1^2 \cup \mathcal{H}_2^2) \cup (\mathcal{H}_1^3 \cup \mathcal{H}_2^3)$$
$$\cong (\bar{A}'_g \cup \tilde{A}'_g) \cup h^2 \cup h^1 \cup (\mathcal{H}_1^2 \cup \mathcal{H}_2^2) \cup (\mathcal{H}_1^3 \cup \mathcal{H}_2^3)$$
$$\cong \bar{A}'_g \cup ((\partial_+ \bar{A}'_g \times [0,1]) \cup h^2) \cup \tilde{A}'_g \cup h^1 \cup (\mathcal{H}_1^2 \cup \mathcal{H}_2^2) \cup (\mathcal{H}_1^3 \cup \mathcal{H}_2^3)$$
$$\cong \bar{A}'_g \cup B'_g \cup ((G_1 \cup G_2) \times [0,1]) \cup \tilde{A}'_g \cup h^1 \cup (\mathcal{H}_1^2 \cup \mathcal{H}_2^2) \cup (\mathcal{H}_1^3 \cup \mathcal{H}_2^3)$$
$$\cong \bar{A}'_g \cup B'_g \cup ((G_1 \cup G_2) \times [0,1]) \cup \tilde{A}'_g \cup h^1 \cup (\mathcal{H}_1^2 \cup \mathcal{H}_2^2) \cup (\mathcal{H}_1^3 \cup \mathcal{H}_2^3)$$
$$\cong \bar{A}'_g \cup (B'_g \cup (G_1 \times [0,1]) \cup \mathcal{H}_1^2 \cup \mathcal{H}_1^3) \cup \tilde{A}'_g$$
$$\qquad \cup ((G_2 \times [0,1]) \cup \tilde{A}'_g \cup h_1) \cup \mathcal{H}_2^2 \cup \mathcal{H}_2^3$$
$$\cong \bar{A}'_g \cup \bar{B}'_g \cup (\tilde{A}'_g \cup A^*_g) \cup ((\partial_+ A^*_g \times [0,1]) \cup \mathcal{H}_2^2 \cup \mathcal{H}_2^3)$$
$$\cong (\bar{A}'_g \cup \bar{B}'_g) \cup (A''_g \cup B''_g).$$

According to this decomposition, we can modify the fork complex $(\mathscr{F}, \rho)$ as in Fig. 5.12 or Fig. 5.13. It is easy to see that for a new complex $(\mathscr{F}^*, \rho^*)$, we have $w(\mathscr{F}^*, \rho^*) < w(\mathscr{F}, \rho)$, a contradiction. □

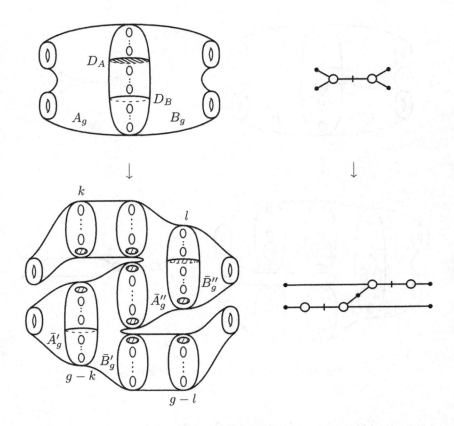

Fig. 5.12   Case 2. The case of irreducible splittings such that each of ∂D_A and ∂D_B is separating in S.

Lemma 5.2.4. *Let t be a tine of $\mathscr{F}$. Either the component $\rho^{-1}(t)$ is incompressible in M or M is ∂-compressible.*

Proof. Suppose that $\rho^{-1}(t)$ is compressible in M for a tine t of $\mathscr{F}$. Let D be a compressing disk of $\rho^{-1}(t)$. Let $\mathcal{T}$ be the union of the tines of $\mathscr{F}$. By an innermost disk argument, we may assume that $D \cap \rho^{-1}(\mathcal{T}) = \partial D$. Let $\mathcal{F}_1$ be the fork containing $\rho(\eta(\partial D; D))$. Note that $\rho^{-1}(t)$ is incompressible in $\rho^{-1}(\mathcal{F}_1)$ (cf. (4) of Remark 2.1.4). Hence there is a fork $\mathcal{F}_2(\neq \mathcal{F}_1)$ attached to the grip, say g, of $\mathcal{F}_1$. Since $D \cap \rho^{-1}(\mathcal{T}) = \partial D$, we have $D \subset \rho^{-1}(\mathcal{F}_1 \cup \mathcal{F}_2)$. Hence D is a ∂-compressing disk of $M' = \rho^{-1}(\mathcal{F}_1) \cup \rho^{-1}(\mathcal{F}_2)$. Hence it fol-

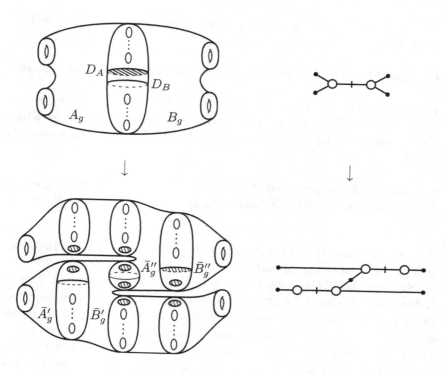

Fig. 5.13 Case 2. The case of reducible splittings such that each of ∂D_A and ∂D_B is separating in S.

lows from (2) of Theorem 3.2.1 and Lemma 3.1.12 that the Heegaard splitting $(\rho^{-1}(\mathcal{F}_1), \rho^{-1}(\mathcal{F}_2); \rho^{-1}(g))$ is either weakly reducible or trivial. It also follows from Proposition 5.2.3 that $(\rho^{-1}(\mathcal{F}_1), \rho^{-1}(\mathcal{F}_2); \rho^{-1}(g))$ is strongly irreducible and hence the splitting must be trivial. Since $\rho^{-1}(\mathcal{F}_1)$ contains $\eta(\partial D; D)$, we see that $\rho^{-1}(\mathcal{F}_1)$ is a trivial compression body and that t is the tine of $\mathcal{F}_1$. Hence by Lemma 5.2.2, we have one of the following: (1) M is a 3-ball, (2) $M \cong \rho^{-1}(t) \times [0, 1]$ and (3) M is ∂-compressible. We suppose that M does not satisfy the condition (3), *i.e.*, M is ∂-incompressible. If M satisfies condition (1), *i.e.*, M is a 3-ball, then it follows from Example 5.1.12 that t is the only tine of $\mathcal{F}$ and that $\rho^{-1}(t)$ is incompressible in M. This contradicts our assumption that $\rho^{-1}(t)$ is compressible in M. If M satisfies condition (2), *i.e.*, $M \cong \rho^{-1}(t) \times [0, 1]$, then $\rho^{-1}(t)$ is incompressible in M, again contradicting our assumption. $\qquad\square$

As a direct consequence of Proposition 5.2.3 and Lemma 5.2.4, we have the following.

Corollary 5.2.5. $(\mathscr{F}, \rho)$ *is strongly irreducible unless M is ∂-compressible.*

Remark 5.2.6. There are strongly irreducible splittings which are not thin. In fact, there are strongly irreducible Heegaard splittings which are not minimal genus (cf. [Casson and Gordon (1986)] and [Kobayashi (1992)]).

Lemma 5.2.7. *Suppose that $\mathscr{F}$ contains a tine. Then there exists a tine t of $\mathscr{F}$ such that $\rho^{-1}(t)$ is a 2-sphere if and only if M is reducible or is a 3-ball.*

Proof. The "only if part" is immediate from Observation 5.2.1. Hence we will give a proof of the "if part".

Suppose that M is reducible or is a 3-ball. If M is a 3-ball, then it follows from Example 5.1.12 that there is exactly a single tine, say t, of $\mathscr{F}$ and $\rho^{-1}(t) = \partial M$ is a 2-sphere. Hence in the remainder of the proof, we suppose that M is reducible. Let $\mathcal{T}$ be the union of the tines of $\mathscr{F}$. Let P be a reducing 2-sphere such that $|P \cap \rho^{-1}(\mathcal{T})|$ is minimal among such all reducing 2-spheres. By an innermost disk argument, we see that $P \cap \rho^{-1}(\mathcal{T}) = \emptyset$. Let $\mathcal{F}_1$ be a fork of $\mathscr{F}$ with $\rho^{-1}(\mathcal{F}_1) \cap P \neq \emptyset$.

Suppose first that there are no forks of $\mathscr{F}$ attaching to the grip of $\mathcal{F}_1$. Then this implies that P is an essential 2-sphere in $\rho^{-1}(\mathcal{F}_1)$.

Suppose next that there is a fork of $\mathscr{F}$, say $\mathcal{F}_2$, other than $\mathcal{F}_1$ which attaches to the grip, say g, of $\mathcal{F}_1$. Note that $\rho^{-1}(\mathcal{F}_1) \cup \rho^{-1}(\mathcal{F}_2)$ contains P. It follows from (1) of Theorem 3.2.1 that $\rho^{-1}(\mathcal{F}_1)$ or $\rho^{-1}(\mathcal{F}_2)$ is reducible, or $(\rho^{-1}(\mathcal{F}_1), \rho^{-1}(\mathcal{F}_2); \rho^{-1}(g))$ is reducible. The latter condition, however, contradicts Proposition 5.2.3. Hence we may assume that $\rho^{-1}(\mathcal{F}_1)$ is reducible, that is, there is a 2-sphere component P_0 of $\partial_-(\rho^{-1}(\mathcal{F}_1))$ (cf. (1) of Remark 2.1.4). This implies that there is a tine t with $\rho^{-1}(t) = P_0$. $\square$

Lemma 5.2.8. *If some $\rho^{-1}(g)$ is a torus, where g is a grip of $\mathscr{F}$, then one of the following holds.*
(1) *M is reducible.*
(2) *M is (a torus) $\times [0, 1]$.*
(3) *M is a solid torus.*
(4) *M is a lens space.*

Proof. Suppose that M does not satisfy the conclusion (1) of Lemma 5.2.8, *i.e.*, M is irreducible. Note that $\rho^{-1}(g)$ may be a boundary component of M. Let $\mathcal{F}$ be a fork such that the grip of $\mathcal{F}$ is g. Set $V = \rho^{-1}(\mathcal{F})$.

If V is trivial, then M is either $\mathbb{T}^2 \times [0, 1]$ or a solid torus by Lemma 5.2.2. Hence conclusion (2) or (3) of Lemma 5.2.8 holds.

If V is non-trivial, then we see that V is a solid torus by Observation 5.2.1 and Example 5.1.12. Suppose further that the conclusion (3) does not hold., *i.e.*, M is not a solid torus. Then there is a fork $\mathcal{F}'(\neq \mathcal{F})$ attached to g. Set $V' = \rho^{-1}(\mathcal{F}')$. If V' is trivial, then it follows from Lemma 5.2.2 that M is a solid torus, a contradiction. If V' is non-trivial, then we see that V' is a solid torus by Observation 5.2.1 and Example 5.1.12. Hence M is a lens space and we have the conclusion (4) of Lemma 5.2.8. □

5.3 Examples

In this section we use some theorems without proofs to obtain generalized Heegaard splittings and associated fork complexes. Let F_g be a connected closed orientable surface of genus g.

5.3.1 $M = F_g \times [0, 1]$.

Set $M = F_g \times [0, 1]$, $A = F_g \times [0, 1/2]$, $B = F_g \times [1/2, 1]$ and $S = F_g \times \{1/2\}$. Clearly, $(A, B; S)$ is a Heegaard splitting of M, and we call this Heegaard splitting the *standard Heegaard splitting of type I*. Let p be a point in F_g. Set

$$A' = \eta((F_g \times \{0\}) \cup (p \times [0, 1]) \cup (F_g \times \{1\}); M),$$

$B' = \text{cl}(M \setminus A')$ and $S' = A' \cap B'$. Then $(A', B'; S')$ is also a Heegaard splitting of M, and we call this splitting the *standard Heegaard splitting of type II*. The proof of the next observation is left to the reader.

Observation 5.3.1. The two Heegaard splittings above are strongly irreducible.

Moreover, the following is proved in [Scharlemann and Thompson (1993), 2.11 Main Theorem].

Theorem 5.3.2 ([Scharlemann and Thompson (1993)]). *Any irreducible Heegaard splitting of $F_g \times [0, 1]$ is standard of type I or II.*

We remark that the fork complexes associated to these Heegaard splittings are illustrated in Fig. 5.14.

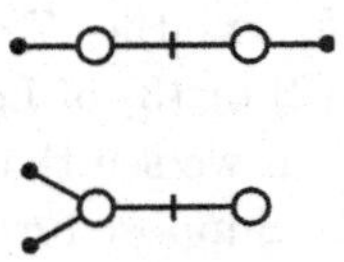

Type I

Type II

Fig. 5.14   The fork complexes associated to irreducible Heegaard splittings of $F_g \times [0,1]$.

5.3.2 $M = F_g \times \mathbb{S}^1$.

Note that $\mathbb{S}^1$ is regarded as $[0,1]/\{0\} \sim \{1\}$. Let p and q be distinct points in F_g. Set

$$A = \mathrm{cl}((F_g \times [0,1/2]) \setminus \eta(p \times [0,1/2]; F_g \times [0,1/2]))$$
$$\cup \eta(q \times [1/2,1]; F_g \times [1/2,1])$$

and

$$B = \mathrm{cl}(M \setminus A)$$
$$= \mathrm{cl}((F_g \times [1/2,1]) \setminus \eta(q \times [1/2,1]; F_g \times [1/2,1]))$$
$$\cup \eta(p \times [0,1/2]; F_g \times [0,1/2]).$$

Note that A and B are handlebodies. Set $S = \partial A \cap \partial B$. Then $(A, B; S)$ is a Heegaard splitting of $M = F_g \times \mathbb{S}^1$ and is called the *standard Heegaard splitting* of $M = F_g \times \mathbb{S}^1$ (cf. Fig. 5.15).

Exercise 5.3.3. Show that the standard Heegaard splitting of $M = F_g \times \mathbb{S}^1$ is weakly reducible.

Theorem 5.3.4 ([Schultens (1993)] Theorem 5.7). *Any   irreducible Heegaard splitting of $F_g \times \mathbb{S}^1$ is the standard splitting.*

We first consider a special case of $M = F_g \times [0,1]$: a 3-torus $\mathbb{T}^3 = \mathbb{T}^2 \times \mathbb{S}^1 (= \mathbb{S}^1 \times \mathbb{S}^1 \times \mathbb{S}^1)$.

It is known that $\mathbb{T}^3$ is obtained from a cube $[0,1] \times [0,1] \times [0,1]$ by attaching corresponding edges and faces as in Fig. 5.16 (a). Set

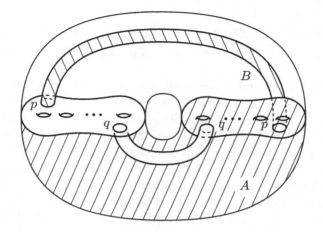

Fig. 5.15   The standard Heegaard splitting of $M = F_g \times \mathbb{S}^1$.

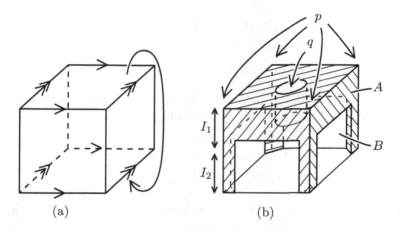

Fig. 5.16   The irreducible Heegaard splitting of $\mathbb{T}^3$.

$$A = \mathrm{cl}((\mathbb{T}^2 \times [0, 1/2]) \setminus \eta(p \times [0, 1/2]; \mathbb{T}^2 \times [0, 1/2]))$$
$$\cup \eta(q \times [1/2, 1]; \mathbb{T}^2 \times [1/2, 1])$$

and

$$B = \mathrm{cl}(\mathbb{T}^3 \setminus A)$$
$$= \mathrm{cl}((\mathbb{T}^2 \times [1/2, 1]) \setminus \eta(q \times [1/2, 1]; \mathbb{T}^2 \times [1/2, 1]))$$
$$\cup \eta(p \times [0, 1/2]; \mathbb{T}^2 \times [0, 1/2]).$$

Then we see that A and B are genus two handlebodies and that it follows from Theorem 5.3.4 that $(A, B; S)$ is the Heegaard splitting of $\mathbb{T}^3$, where $S = \partial A = \partial B$ (cf. Fig. 5.16 (b)). Set $h^1 = \eta(q \times [1/2, 1]; \mathbb{T}^2 \times [1/2, 1])$ and $h^2 = \eta(p \times [0, 1/2]; \mathbb{T}^2 \times [0, 1/2])$. Note that h^1 (h^2 resp.) can be regarded as a 1-handle (2-handle resp.) in a handle decomposition of $\mathbb{T}^3$ obtained from the Heegaard splitting $(A, B; S)$. Since $h^1 \cap h^2 = \emptyset$, we can perform a weak reduction to obtain a generalized Heegaard splitting. We give a concrete description of the generalized Heegaard splitting in the following. First, set $A_1 = \mathrm{cl}(\mathbb{T}^2 \times [0, 1/2] \setminus h^1)$ and $B_2 = \mathrm{cl}(\mathbb{T}^2 \times [1/2, 1] \setminus h^2)$. That is, A_1 is obtained from A by removing the 1-handle h^1 and B_2 is obtained from B by removing the 2-handle h^2. Then we have

$$\begin{aligned}
\mathbb{T}^3 &= A \cup B \\
&= A_1 \cup h^1 \cup h^2 \cup B_2 \\
&\cong A_1 \cup (\partial A_1 \times [0, 1]) \cup h^1 \cup h^2 \cup B_2 \\
&= A_1 \cup ((\partial A_1 \times [0, 1]) \cup h^2) \cup h^1 \cup B_2.
\end{aligned}$$

Set $B_1 = (\partial A_1 \times [0, 1]) \cup h^2$. Then B_1 is a compression body such that $\partial_+ B_1 = \partial A_1$ and $\partial_- B$ consists of two tori. Hence we have

$$\begin{aligned}
\mathbb{T}^3 &\cong A_1 \cup B_1 \cup h^1 \cup B_2 \\
&\cong A_1 \cup B_1 \cup ((\partial_- B_1 \times [0, 1]) \cup h^1) \cup B_2.
\end{aligned}$$

Set $A_2 = (\partial_- B_1 \times [0, 1]) \cup h^1$. Then A_2 is a compression body such that $\partial_+ A_2 = \partial B_2$ and $\partial_- A_2 = \partial_- B_1$. Hence we have

$$\mathbb{T}^3 = (A_1 \cup B_1) \cup (A_2 \cup B_2).$$

This together with the fork complex as in Fig. 5.17 gives a generalized Heegaard splitting.

Exercise 5.3.5. Show that this is the only fork complex associated with a generalized Heegaard splitting of $\mathbb{T}^3$ via weak reduction.

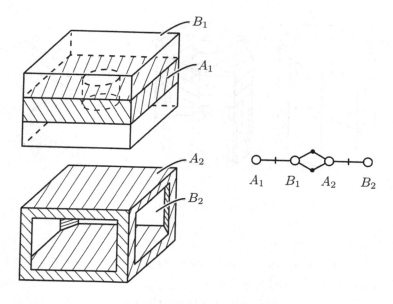

Fig. 5.17 A generalized Heegaard splitting of $\mathbb{T}^3$ and its fork complex.

Exercise 5.3.6. Show that the above generalized Heegaard splitting of $\mathbb{T}^3$ is strongly irreducible.

We next consider $M = F_2 \times \mathbb{S}^1$ as another example of $F_g \times [0,1]$. Then M is obtained from (an octagon)$\times[0,1]$ by attaching corresponding edges and faces as in Fig. 5.18 (a). Set

$$A = \mathrm{cl}((F_g \times [0,1/2]) \setminus \eta(p \times [0,1/2]; F_g \times [0,1/2]))$$
$$\cup \eta(q \times [1/2,1]; F_g \times [1/2,1])$$

and

$$B = \mathrm{cl}(M \setminus A)$$
$$= \mathrm{cl}((F_g \times [1/2,1]) \setminus \eta(q \times [1/2,1]; F_g \times [1/2,1]))$$
$$\cup \eta(p \times [0,1/2]; F_g \times [0,1/2]).$$

Then it follows from Theorem 5.3.4 that we obtain the Heegaard splitting $M = A \cup B$ (see Fig. 5.18 (b)). As described in case of $M = \mathbb{T}^3$, we can

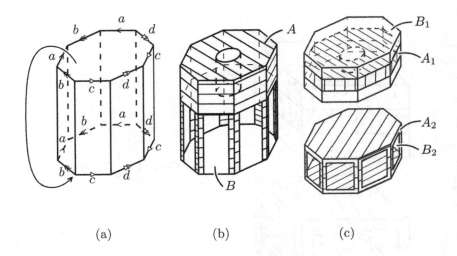

(a)   (b)   (c)

Fig. 5.18   The irreducible Heegaard splitting of $F_2 \times \mathbb{S}^1$.

perform a weak reduction and we obtain the same fork complex as that illustrated in Fig. 5.17. In this case each of A_1 and B_2 is a handlebody of genus four and each of A_2 and B_1 is a compression body with $\partial_+ A_2 = \partial B_2$, $\partial A_1 = \partial_+ B_1$ and $\partial_- A_2 = \partial_- B_1$.

For the Heegaard splitting $A \cup B$ of M, we can find another weak reduction as follows. Recall that $M = P_8 \times [0,1]/ \sim$, where P_8 is an octagon (cf. Fig. 5.18). Then there is a handle decomposition

$$M = h^0 \cup h_a^1 \cup h_b^1 \cup h_c^1 \cup h_d^1 \cup h_e^1 \cup h_a^2 \cup h_b^2 \cup h_c^2 \cup h_d^2 \cup h_e^2 \cup h^3,$$

where a 0-handle h^0 corresponds to a vertex of P_8, a 1-handle h_a^1 (h_b^1, h_c^1, h_d^1 and h_e^1 resp.) corresponds to a (b, c, d and e resp.) in P_8, a 2-handle h_a^2 (h_b^2, h_c^2 and h_d^2 resp.) corresponds to the face bounded by $eae^{-1}a^{-1}$ ($ebe^{-1}b^{-1}$, $ece^{-1}c^{-1}$ and $ede^{-1}d^{-1}$ resp.) in $\partial P_8 \times [0,1]$, a 2-handle h_e^2 corresponds to the face bounded by $aba^{-1}b^{-1}cdc^{-1}d^{-1}$ in P_8 and a 3-handle h^3 corresponds to the vertex in the interior of $P_8 \times [0,1]$. Set $A_1 = h^0 \cup h_a^1 \cup h_e^1$. Then A_1 is a genus two handlebody and we have

$$M = A_1 \cup h_b^1 \cup h_c^1 \cup h_d^1 \cup h_a^2 \cup h_b^2 \cup h_c^2 \cup h_d^2 \cup h_e^2 \cup h^3$$
$$\cong A_1 \cup (\partial A_1 \times [0,1]) \cup h_b^1 \cup h_c^1 \cup h_d^1 \cup h_a^2 \cup h_b^2 \cup h_c^2 \cup h_d^2 \cup h_e^2 \cup h^3$$
$$= A_1 \cup \big((\partial A_1 \times [0,1]) \cup h_a^2\big) \cup h_b^1 \cup h_c^1 \cup h_d^1 \cup h_b^2 \cup h_c^2 \cup h_d^2 \cup h_e^2 \cup h^3.$$

Set $B_1 = (\partial A_1 \times [0,1]) \cup h_a^2$. Then B_1 is a compression body such that $\partial_+ B_1 = \partial A_1$ and $\partial_- B_1$ consists of two tori. Then we have

$$M \cong A_1 \cup B_1 \cup h_b^1 \cup h_c^1 \cup h_d^1 \cup \cup h_b^2 \cup h_c^2 \cup h_d^2 \cup h_e^2 \cup h^3$$
$$\cong A_1 \cup B_1 \cup (\partial_- B_1 \times [0,1]) \cup h_b^1 \cup h_c^1 \cup h_d^1 \cup \cup h_b^2 \cup h_c^2 \cup h_d^2 \cup h_e^2 \cup h^3$$
$$= A_1 \cup B_1 \cup ((\partial_- B_1 \times [0,1]) \cup h_b^1) \cup h_c^1 \cup h_d^1 \cup \cup h_b^2 \cup h_c^2 \cup h_d^2 \cup h_e^2 \cup h^3.$$

Set $A_2 = (\partial_- B_1 \times [0,1]) \cup h_b^1$. Then A_2 is a compression body such that $\partial_+ A_2$ is a closed surface of genus two and $\partial_- A_2 = \partial_- B_1$. Then we have

$$M \cong A_1 \cup B_1 \cup A_2 \cup h_c^1 \cup h_d^1 \cup \cup h_b^2 \cup h_c^2 \cup h_d^2 \cup h_e^2 \cup h^3$$
$$\cong A_1 \cup B_1 \cup A_2 \cup (\partial_+ A_2 \times [0,1]) \cup h_c^1 \cup h_d^1 \cup \cup h_b^2 \cup h_c^2 \cup h_d^2 \cup h_e^2 \cup h^3$$
$$= A_1 \cup B_1 \cup A_2 \cup ((\partial_+ A_2 \times [0,1]) \cup h_b^2) \cup h_c^1 \cup h_d^1 \cup h_c^2 \cup h_d^2 \cup h_e^2 \cup h^3.$$

Set $B_2 = (\partial_+ A_2 \times [0,1]) \cup h_b^2$. Then B_2 is a compression body such that $\partial_+ B_2 = \partial_+ A_2$ and $\partial_- B_2$ consists of a torus. Then we have

$$M \cong A_1 \cup B_1 \cup A_2 \cup B_2 \cup h_c^1 \cup h_d^1 \cup h_c^2 \cup h_d^2 \cup h_e^2 \cup h^3$$
$$\cong A_1 \cup B_1 \cup A_2 \cup B_2 \cup (\partial_- B_2 \times [0,1]) \cup h_c^1 \cup h_d^1 \cup h_c^2 \cup h_d^2 \cup h_e^2 \cup h^3$$
$$= A_1 \cup B_1 \cup A_2 \cup B_2 ((\partial_- B_2 \times [0,1]) \cup h_c^1) \cup h_d^1 \cup h_c^2 \cup h_d^2 \cup h_e^2 \cup h^3.$$

Set $A_3 = (\partial_- B_2 \times [0,1]) \cup h_c^1$. Then A_3 is a compression body such that $\partial_+ A_3$ is a closed surface of genus two and $\partial_- A_3 = \partial_- B_2$. Then we have

$$M \cong A_1 \cup B_1 \cup A_2 \cup B_2 \cup A_3 \cup h_d^1 \cup h_c^2 \cup h_d^2 \cup h_e^2 \cup h^3$$
$$\cong A_1 \cup B_1 \cup A_2 \cup B_2 \cup A_3 \cup (\partial_+ A_3 \times [0,1]) \cup h_d^1 \cup h_c^2 \cup h_d^2 \cup h_e^2 \cup h^3$$
$$= A_1 \cup B_1 \cup A_2 \cup B_2 \cup A_3 \cup ((\partial_+ A_3 \times [0,1]) \cup h_c^2) \cup h_d^1 \cup h_d^2 \cup h_e^2 \cup h^3.$$

Set $B_3 = (\partial_+ A_3 \times [0,1]) \cup h_c^2$. Then B_3 is a compression body such that $\partial_+ B_3 = \partial_+ A_3$ and $\partial_- B_3$ consists of two tori. Then we have

$$M \cong A_1 \cup B_1 \cup A_2 \cup B_2 \cup A_3 \cup B_3 \cup h_d^1 \cup h_d^2 \cup h_e^2 \cup h^3$$
$$\cong A_1 \cup B_1 \cup A_2 \cup B_2 \cup A_3 \cup B_3 \cup (\partial_- B_3 \times [0,1]) \cup h_d^1 \cup h_d^2 \cup h_e^2 \cup h^3$$
$$= A_1 \cup B_1 \cup A_2 \cup B_2 \cup A_3 \cup B_3 \cup ((\partial_- B_3 \times [0,1]) \cup h_d^1) \cup h_d^2 \cup h_e^2 \cup h^3.$$

Set $A_4 = (\partial_- B_4 \times [0,1]) \cup h_d^1$. Then A_4 is a compression body such that $\partial_+ A_4$ is a closed surface of genus two and $\partial_- A_4 = \partial_- B_3$. Then we have

$$M \cong A_1 \cup B_1 \cup A_2 \cup B_2 \cup A_3 \cup B_3 \cup A_4 \cup h_d^2 \cup h_e^2 \cup h^3$$
$$\cong A_1 \cup B_1 \cup A_2 \cup B_2 \cup A_3 \cup B_3 \cup A_4 \cup (\partial_+ A_4 \times [0,1]) \cup h_d^2 \cup h_e^2 \cup h^3.$$

Set $B_4 = (\partial_+ A_4 \times [0,1]) \cup h_d^2 \cup h_e^2 \cup h^3$. Then B_4 is a genus two handlebody such that $\partial B_4 = \partial_+ A_4$. Therefore we have the following decomposition:

$$F_2 \times \mathbb{S}^1 = (A_1 \cup B_1) \cup (A_2 \cup B_2) \cup (A_3 \cup B_3) \cup (A_4 \cup B_4).$$

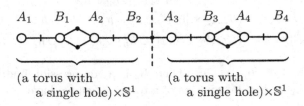

Fig. 5.19

This together with the fork complex as in Fig. 5.19 gives a generalized Heegaard splitting. We remark that $(A_1 \cup B_1) \cup (A_2 \cup B_2)$ $((A_3 \cup B_3) \cup (A_4 \cup B_4)$ resp.) composes a (a torus with a single hole) $\times \mathbb{S}^1$.

By changing the attaching order of h_b^1, h_c^1 and h_d^1, we can obtain two more strongly irreducible generalized Heegaard splittings via weak reduction (cf. Fig. 5.20).

Exercise 5.3.7. Show that these are the only fork complexes associated with a generalized Heegaard splitting of $F_2 \times \mathbb{S}^1$ via weak reduction.

Remark 5.3.8. The fork complexes associated with distinct weak reduction of a Heegaard splitting need not to be homotopic.

5.4 Amalgamation

One can easily construct a generalized Heegaard splitting of a 3-manifold. Let M_1 and M_2 be 3-manifolds with boundary such that a boundary component, say F_1, of M_1 is homeomorphic to a boundary component, say F_2,

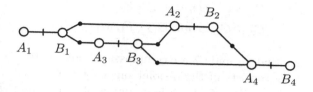

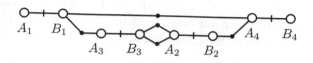

Fig. 5.20

of M_2. Let M be a 3-manifold obtained from M_1 and M_2 by identifying F_1 with F_2, and let F be the image of F_1 (and hence of F_2) in M. For each $i = 1$ and 2, let $(C_{i1}, C_{i2}; S_i)$ be a Heegaard splitting of M_i. We may assume that $F_1 \subset \partial_- C_{12}$ and $F_2 \subset \partial_- C_{21}$. Then

$$M = (C_{11} \cup C_{12}) \cup_F (C_{21} \cup C_{22})$$

gives a generalized Heegaard splitting of M.

There is a way, called *amalgamation*, to obtain a Heegaard splitting of M from two Heegaard splittings $(C_{i1}, C_{i2}; S_i)$ of M_i $(i = 1, 2)$. This was originally introduced by the second author. See [Schultens (1993)]. Roughly speaking, amalgamation is the inverse procedure of weak reduction. We describe more precisely how to amalgamate these two Heegaard splittings.

Let Σ_{12} and Σ_{21} be spines of C_{12} and C_{21} respectively. We regard C_{12} as a compression body obtained from $\eta(\partial_- C_{12}; M_1)$ by attaching 1-handles, say $\mathcal{H}_{12}^1$, induced from Σ_{12}. Similarly, we regard C_{21} as a compression body obtained from $\eta(\partial_- C_{21}; M_2)$ by attaching 1-handles, say $\mathcal{H}_{21}^1$, induced from Σ_{21}. Note that $\eta(F_1; M_1) \cup_F \eta(F_2; M_2)$ is homeomorphic to $F \times [0, 1]$ such that (i) $F \times \{0\} = \partial \eta(F_1; M_1) \setminus F_1$, (ii) $F \times \{1/2\} = F$, and (iii) $F \times \{1\} = \partial \eta(F_2; M_2) \setminus F_2$.

We now collapse $F \times [0, 1]$, *i.e.*, identify $F \times \{t\}$ with $F \times \{1/2\}$ for any $t \in [0, 1]$ so that the attaching spheres of $\mathcal{H}_{12}^1$ are disjoint from those of $\mathcal{H}_{21}^1$. Note that the resulting manifold is homeomorphic to the original manifold M. Set

$$C_1 = C_{11} \cup \eta(\partial_- C_{21} \setminus F_2; M_2) \cup \mathcal{H}_{21}^1,$$
$$C_2 = C_{22} \cup \eta(\partial_- C_{12} \setminus F_1; M_1) \cup \mathcal{H}_{12}^1.$$

It is easy to see that C_1 and C_2 are compression bodies with $\partial_+ C_1 = \partial_+ C_2$. Note that $\partial_- C_1$ consists of the disjoint union of $\partial_- C_{11}$ and $\partial_- C_{21} \setminus F_2$. Similarly, $\partial_- C_2$ consists of the disjoint union of $\partial_- C_{22}$ and $\partial_- C_{12} \setminus F_1$. Hence $(C_1, C_2; S)$ is a Heegaard splitting of $(M; \partial_1 M, \partial_2 M)$, where $S = \partial_+ C_1 = \partial_+ C_2$, $\partial_1 M = \partial_- C_1$ and $\partial_2 M = \partial_- C_2$. The Heegaard splitting $(C_1, C_2; S)$ is called the *amalgamation* of $(C_{11}, C_{12}; S_1)$ and $(C_{21}, C_{22}; S_2)$ along F. An example of amalgamation is illustrated in Fig. 5.21.

Remark 5.4.1. Let $(C_1, C_2; S)$ be the amalgamation of two Heegaard splittings $(C_{11}, C_{12}; S_1)$ and $(C_{21}, C_{22}; S_2)$ along F as above. Then the following holds:

$$\text{genus}(S) = \text{genus}(S_1) + \text{genus}(S_2) - \text{genus}(F).$$

Exercise 5.4.2. Show Remark 5.4.1.

We can similarly amalgamate two Heegaard splittings along possibly disconnected surfaces and also amalgamate more than two Heegaard splittings. See [Schultens (1993)] and [Derby-Talbot (2009)].

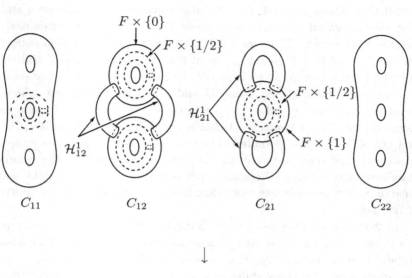

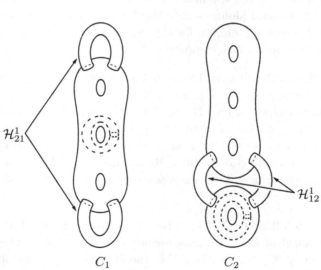

Fig. 5.21   Amalgamating two Heegaard splittings along a surface F.

5.5 Boundary stabilization and the stabilization problem

Recall that Theorem 4.2.4, *i.e.*, the Reidemeister-Singer Theorem states that two Heegaard splittings of a closed 3-manifold can be stabilized to become isotopic (cf. Theorem 4.2.4). The *stabilization problem* concerns the question of how high the genus of the stabilized Heegaard splittings must be for this isotopy to be achieved. In 1996, the second author showed that for any Seifert fibered space M and any pair of Heegaard splittings $(C_1, C_2; S)$ of genus g and $(C_1', C_2'; S')$ of genus g' with $g \geq g'$, there is a Heegaard splitting of genus $g + 1$ that is a stabilization of both $(C_1, C_2; S)$ and $(C_1', C_2'; S')$. See [Schultens (1996)]. For many years, this example was considered representative, *i.e.*, it was conjectured that the common stabilization of a pair of Heegaard splittings of a given 3-manifold would have to have genus only one more than the higher genus Heegaard splitting of the pair.

In 2008, Hass, Thompson and Thurston exhibited examples for every $g > 0$, pairs of Heegaard splittings of genus g where the genus of a common stabilization needs to be $2g$. It is conjectured that these examples provide the upper bound for the stabilization problem. See [Hass *et al.* (2009)]. In [Johnson (2010)] and [Johnson (2011)], Johnson obtained related results. The main theorem in [Johnson (2011)] uses a construction that lends itself to an examination via fork complexes.

Remark 5.5.1. Recall that Theorem 5.3.4 states that every unstabilized Heegaard splitting of $F \times [0, 1]$ is one of the standard Heegaard splittings, *i.e.*, is either of type I or type II (cf. [Scharlemann and Thompson (1993)]). Note that the partition of boundary components of a Heegaard splitting does not change under stabilization. Thus these two Heegaard splittings do not admit a common stabilization. Recall again that Theorem 4.2.4, *i.e.*, the Reidemeister-Singer Theorem was stated for and applies only to closed 3-manifolds.

Definition 5.5.2. Let $(C_1, C_2; S)$ be a Heegaard splitting of M and let F be a component of ∂M. We may assume that $F \subset \partial_- C_2$. The *boundary stabilization* of $(C_1, C_2; S)$ *along* F is the Heegaard splitting obtained by amalgamating $(C_1, C_2; S)$ with a type II Heegaard splitting of $F \times [0, 1]$.

The fork complex of a boundary stabilization before the amalgamation is as in Fig. 5.22. Note that the boundary stabilization has genus one more than the genus of the original Heegaard surface.

Johnson uses this idea in [Johnson (2010)]. He constructs 3-manifolds with Heegaard splittings of genus g satisfying numerous technical requirements. He then compares these Heegaard splittings with their boundary stabilization and finds that their common stabilization must have genus at least $2g$.

He also constructs pairs of 3-manifolds M_1 and M_2 with Heegaard splittings $(C_{11}, C_{12}; S_1)$ and $(C_{21}, C_{22}; S_2)$ respectively. The manifolds are chosen so that each of M_1 and M_2 contains a boundary component F_1 and F_2 respectively such that F_1 is homeomorphic to F_2. Let M be a 3-manifold obtained from M_1 and M_2 by attaching F_1 to F_2, and let F be the image of F_1 and F_2 in M. Johnson then constructs two non-isotopic Heegaard splittings of M from $(C_{11}, C_{12}; S_1)$ and $(C_{21}, C_{22}; S_2)$. The first is simply the amalgamation of $(C_{11}, C_{12}; S_1)$ and $(C_{21}, C_{22}; S_2)$ along F. The second is the amalgamation along F_1 of $(C_{11}, C_{12}; S_1)$ and the boundary stabilization of $(C_{21}, C_{22}; S_2)$ along F_2. The fork complexes are described in Fig 5.23.

In [Johnson (2010)], each M_i $(i = 1, 2)$ is assumed to have a single boundary component F_i which is homeomorphic to a torus, and is assumed to have a "sufficiently complicated" Heegaard splitting $(C_{i1}, C_{i2}; S_i)$ of genus g. Then it is clear from Remark 5.4.1 that the first amalgamation described above is of genus $2g - 1$ and the second is of genus $2g$. A main theorem in [Johnson (2010)] states that for these Heegaard splittings, a common stabilization must have genus at least $3g - 1$. For a generalization of this result, see Johnson and Tomova [Johnson and Tomova (2011)]. For a refinement, see [Takao (2011)].

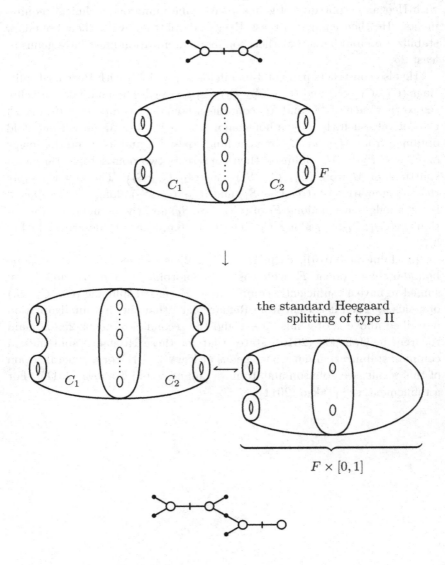

Fig. 5.22   The fork complex of a boundary stabilization before the amalgamation.

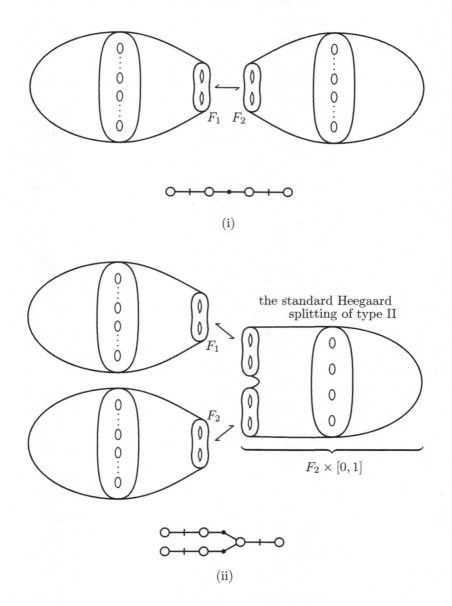

Fig. 5.23 (i) The first Heegaard splitting of M is obtained by amalgamating $(C_{11}, C_{12}; S_1)$ and $(C_{21}, C_{22}; S_2)$. (ii) The second Heegaard splitting of M is obtained by the amalgamation along F_1 of $(C_{11}, C_{12}; S_1)$ and the boundary stabilization of $(C_{21}, C_{22}; S_2)$ along F_2.

Bibliography

Adams, C. C. (1994). *The knot book. An elementary introduction to the mathematical theory of knots.* (W. H. Freeman and Company, New York), ISBN 0-7167-2393-X.

Akbulut, S. and McCarthy, J. D. (1990). *Casson's invariant for oriented homology 3-spheres, Mathematical Notes,* Vol. 36 (Princeton University Press, Princeton, NJ), ISBN 978-0691085630, an exposition.

Alexander, J. W. (1924). On the subdivision of 3-space by a polyhedron, *Proc. Natl. Acad. Sci. USA* **10**, No. 1, pp. 6–8, http://www.ncbi.nlm.nih.gov/pmc/articles/PMC1085499/.

Bonahon, F. and Otal, J.-P. (1983). Scindements de Heegaard des espaces lenticulaires, *Ann. Sci. École Norm. Sup. (4)* **16**, No. 3, pp. 451–466, http://www.numdam.org/item?id=ASENS_1983_4_16_3_451_0.

Brody, E. J. (1960). The topological classification of the lens spaces, *Ann. of Math.* **71**, No. 1, pp. 163–184, doi:10.2307/1969884, http://www.jstor.org/stable/1969884.

Casson, A. J. and Gordon, C. M. (1986). Manifolds with irreducible Heegaard splittings of arbitrarily high genus, (unpublished).

Casson, A. J. and Gordon, C. M. (1987). Reducing Heegaard splittings, *Topology Appl.* **27**, No. 3, pp. 275–283, doi:10.1016/0166-8641(87)90092-7, http://dx.doi.org/10.1016/0166-8641(87)90092-7.

Derby-Talbot, R. (2009). Stabilization, amalgamation and curves of intersection of Heegaard splittings, *Algebr. Geom. Topol.* **9**, No. 2, pp. 811–832, doi:10.2140/agt.2009.9.811, http://dx.doi.org/10.2140/agt.2009.9.811.

do Carmo, M. P. a. (1992). *Riemannian geometry,* Mathematics: Theory & Applications (Birkhäuser Boston, Inc., Boston, MA), ISBN 978-0817634902, translated from the second Portuguese edition by Francis Flaherty.

Freedman, M. H. (1982). The topology of four-dimensional manifolds, *J. Differential Geom.* **17**, No. 3, pp. 357–453, https://projecteuclid.org/euclid.jdg/1214437136.

Gabai, D. (1987). Foliations and the topology of 3-manifolds III, *J. Differential Geom.* **26**, No. 3, pp. 479–536, http://projecteuclid.org/euclid.jdg/1214441488.

Guillemin, V. and Pollack, A. (2010). *Differential topology* (AMS Chelsea Publishing, Providence, RI), ISBN 978-0-8218-5193-7, reprint of the 1974 original published by Prentice Hall Inc.

Haken, W. (1968). *Some results on surfaces in 3-manifolds*, Studies in Modern Topology, Vol. 5 (Math. Assoc. Amer., Washington D.C.), pp. 39–98.

Hass, J., Thompson, A., and Thurston, W. (2009). Stabilization of Heegaard splittings, *Geom. Topol.* **13**, No. 4, pp. 2029–2050, http://dx.doi.org/10.2140/gt.2009.13.2029.

Hatcher, A. (1980's). Notes on basic 3-manifold topology, Preprint.

Hempel, J. (2004). *3-manifolds* (AMS Chelsea Publishing, Providence, RI), ISBN 978-0-8218-3695-8, reprint of the 1976 original.

Jaco, W. (1980). *Lectures on three-manifold topology*, CBMS Regional Conference Series in Mathematics, Vol. 43 (American Mathematical Society, Providence, R.I.), ISBN 978-0-8218-1693-6.

Johannson, K. (1991). On surfaces and Heegaard surfaces, *Trans. Amer. Math. Soc.* **325**, No. 2, pp. 573–591, doi:10.1090/S0002-9947-1991-1064268-2, http://dx.doi.org/10.2307/2001640.

Johnson, J. (2010). Bounding the stable genera of Heegaard splittings from below, *J. Topol.* **3**, No. 3, pp. 668–690, http://dx.doi.org/10.1112/jtopol/jtq021.

Johnson, J. (2011). *Layered models for closed 3-manifolds*, Contemp. Math., Vol. 560 (Amer. Math. Soc., Providence, RI), http://dx.doi.org/10.1090/conm/560/11090.

Johnson, J. and Tomova, M. (2011). Flipping bridge surfaces and bounds on the stable bridge number, *Algebr. Geom. Topol.* **11**, No. 4, pp. 1987–2005, http://dx.doi.org/10.2140/agt.2011.11.1987.

Kirby, R. C. and Siebenmann, L. C. (1969). On the triangulation of manifolds and the Hauptvermutung, *Bull. Amer. Math. Soc.* **75**, No. 4, pp. 742–749, doi:10.1090/S0002-9904-1969-12271-8, http://dx.doi.org/10.1090/S0002-9904-1969-12271-8.

Kobayashi, T. (1992). A construction of 3-manifolds whose homeomorphism classes of Heegaard splittings have polynomial growth, *Osaka J. Math.* **29**, No. 4, pp. 653–674, http://projecteuclid.org/euclid.ojm/1200784083.

Kobayashi, T. (1994). A construction of arbitrarily high degeneration of tunnel numbers of knots under connected sum, *J. Knot Theory Ramifications* **3**, No. 2, pp. 179–186, doi:10.1142/S0218216594000137, http://dx.doi.org/10.1142/S0218216594000137.

Kobayashi, T. (2001). Heegaard splittings of exteriors of two bridge knots, *Geom. Topol.* **5**, pp. 609–650 (electronic), doi:10.2140/gt.2001.5.609, http://dx.doi.org/10.2140/gt.2001.5.609.

Kobayashi, T. (2003). Scharlemann-Thompson untelescoping of Heegaard splittings is finer than Casson-Gordon's, *J. Knot Theory Ramifications* **12**, No. 7, pp. 877–891, doi:10.1142/S0218216503002810, http://dx.doi.org/10.1142/S0218216503002810.

Kobayashi, T. and Saito, T. (2010). Destabilizing Heegaard splittings of knot exteriors, *Topology Appl.* **157**, no. 1, pp. 202–212, doi:10.1016/j.topol.2009.

04.059, http://dx.doi.org/10.1016/j.topol.2009.04.059.

Lei, F. (2000). On stability of Heegaard splittings, *Math. Proc. Cambridge Philos. Soc.* **129**, No. 1, pp. 55–57, doi:10.1017/S0305004100004461, http://dx.doi.org/10.1017/S0305004100004461.

Massey, W. S. (1977). *Algebraic topology: an introduction*, Graduate Texts in Mathematics, Vol. 56 (Springer-Verlag New York, New York), ISBN 978-0-387-90271-5, reprint of the 1967 edition published by Harcourt, Brace & World Inc.

Milnor, J. W. (1963). *Morse theory*, Ann. of Math. Studies, Vol. 51 (Princeton University Press, Princeton, NJ), based on lecture notes by M. Spivak and R. Wells.

Milnor, J. W. (1997). *Topology from the differentiable viewpoint*, Princeton Landmarks in Mathematics (Princeton University Press, Princeton, NJ), ISBN 978-0691048338, based on notes by David W. Weaver, Revised reprint of the 1965 original.

Moise, E. E. (1952). Affine structures in 3-manifolds. V. The triangulation theorem and Hauptvermutung, *Ann. of Math. (2)* **56**, No. 1, pp. 96–114, doi: 10.2307/1969769, http://dx.doi.org/10.2307/1969769.

Morimoto, K. (1995a). Characterization of tunnel number two knots which have the property "2 + 1 = 2", *Topology Appl.* **64**, No. 2, pp. 165–176, doi:10.1016/0166-8641(94)00096-L, http://dx.doi.org/10.1016/0166-8641(94)00096-L.

Morimoto, K. (1995b). There are knots whose tunnel numbers go down under connected sum, *Proc. Amer. Math. Soc.* **123**, No. 11, pp. 3527–3532, doi: 10.2307/2161103, http://dx.doi.org/10.2307/2161103.

Morimoto, K. (2000). On the super additivity of tunnel number of knots, *Math. Ann.* **317**, No. 3, pp. 489–508, doi:10.1007/PL00004411, http://dx.doi.org/10.1007/PL00004411.

Morimoto, K. and Schultens, J. (2000). Tunnel numbers of small knots do not go down under connected sum, *Proc. Amer. Math. Soc.* **128**, No. 1, pp. 269–278, doi:10.1090/S0002-9939-99-05160-6, http://dx.doi.org/10.1090/S0002-9939-99-05160-6.

Otal, J.-P. (1991). Sur les scindements de Heegaard de la sphère S^3, *Topology* **30**, No. 2, pp. 249–257, doi:10.1016/0040-9383(91)90011-R, http://dx.doi.org/10.1016/0040-9383(91)90011-R.

Radó, T. (1925). Über den begriff der riemannschen fläche, *Acta Sci. Math. (Szeged)* **2**, No. 2-2, pp. 101–121.

Reidemeister, K. (1933). Zur dreidimensionalen Topologie, *Abh. Math. Semin. Univ. Hambg.* **9**, No. 1, pp. 189–194, doi:10.1007/BF02940644, http://dx.doi.org/10.1007/BF02940644.

Reidemeister, K. (1935). Homotopieringe und Linsenräume, *Abh. Math. Semin. Univ. Hambg.* **11**, No. 1, pp. 102–109, doi:10.1007/BF02940717, http://dx.doi.org/10.1007/BF02940717.

Rolfsen, D. (1990). *Knots and links*, Mathematics Lecture Series, Vol. 7 (Publish or Perish, Inc., Houston, TX), ISBN 978-0821834367, corrected reprint of the 1976 original.

Rubinstein, H. and Scharlemann, M. (1996). Comparing Heegaard splittings of non-Haken 3-manifolds, *Topology* **35**, No. 4, pp. 1005–1026, doi:10.1016/0040-9383(95)00055-0, http://dx.doi.org/10.1016/0040-9383(95)00055-0.

Scharlemann, M. and Thompson, A. (1993). Heegaard splittings of (surface)$\times I$ are standard, *Math. Ann.* **295**, No. 1, pp. 549–564, doi:10.1007/BF01444902, http://dx.doi.org/10.1007/BF01444902.

Scharlemann, M. and Thompson, A. (1994a). Thin position and Heegaard splittings of the 3-sphere, *J. Differential Geom.* **39**, No. 2, pp. 343–357, http://projecteuclid.org/euclid.jdg/1214454875.

Scharlemann, M. and Thompson, A. (1994b). *Thin position for 3-manifolds*, *Contemp. Math.*, Vol. 164 (Amer. Math. Soc., Providence, RI), doi: 10.1090/conm/164/01596, http://dx.doi.org/10.1090/conm/164/01596.

Schoenflies, A. M. (1906). Beiträge zur theorie der punktmengen iii, *Math. Ann.* **62**, No. 2, pp. 286–328, doi:10.1007/BF01449982, http://dx.doi.org/10.1007/BF01449982.

Schubert, H. (1954). Über eine numerische Knoteninvariante, *Math. Z.* **61**, pp. 245–288.

Schubert, H. (1956). Knoten mit zwei Brücken, *Math. Z.* **65**, pp. 133–170.

Schultens, J. (1993). The classification of Heegaard splittings for (compact orientable surfaces)$\times S^1$, *Proc. London Math. Soc. (3)* **67**, No. 2, pp. 425–448, doi:10.1112/plms/s3-67.2.425, http://dx.doi.org/10.1112/plms/s3-67.2.425.

Schultens, J. (1996). The stabilization problem for Heegaard splittings of Seifert fibered spaces, *Topology Appl.* **73**, No. 2, pp. 133–139, http://dx.doi.org/10.1016/0166-8641(96)00030-2.

Schultens, J. (2001/02). Genus 2 closed hyperbolic 3-manifolds of arbitrarily large volume, *Topology Proc.* **26**, No. 1, pp. 317–321.

Schultens, J. (2003). Additivity of bridge numbers of knots, *Math. Proc. Cambridge Philos. Soc.* **135**, No. 3, pp. 539–544, doi:10.1017/S0305004103006832, http://dx.doi.org/10.1017/S0305004103006832.

Sedgwick, E. (1997). An infinite collection of Heegaard splittings that are equivalent after one stabilization, *Math. Ann.* **308**, No. 1, pp. 65–72, doi:10.1007/s002080050064, http://dx.doi.org/10.1007/s002080050064.

Singer, I. M. and Thorpe, J. A. (1976). *Lecture notes on elementary topology and geometry*, Undergraduate Texts in Mathematics (Springer-Verlag New York, New York), ISBN 978-1-4615-7347-0, reprint of the 1967 edition published by Scott, Foresman & Co.

Singer, J. (1933). Three-dimensional manifolds and their Heegaard diagrams, *Trans. Amer. Math. Soc.* **35**, pp. 88–111, doi:10.1090/S0002-9947-1933-1501673-5, http://dx.doi.org/10.1090/S0002-9947-1933-1501673-5.

Takao, K. (2011). A refinement of Johnson's bounding for the stable genera of Heegaard splittings, *Osaka J. Math.* **48**, No. 1, pp. 251–268, http://projecteuclid.org/euclid.ojm/1300802713.

Waldhausen, F. (1968). Heegaard-Zerlegungen der 3-Sphäre, *Topology* **7**, No. 2, pp. 195–203, doi:10.1016/0040-9383(68)90027-X, http://dx.doi.org/10.1016/0040-9383(68)90027-X.

Index